Civil engineering insurance and bonding

CIVIL ENGINEERING MANAGEMENT

Series editor S.H. Wearne, BSc(Eng), PhD, FICE, FBIM, Consultant, Director of Institution courses and in-company training

Editorial panel D.E. Neale, CEng, FICE; D.P. Maguire, BSc, FICE; and D.J. Ricketts, BSc; B.A.O. Hewett, BSc(Eng), MSc, FICE; J.V. Tagg, CEng, FICE

Forthcoming titles in the series

Construction planning, R.H. Neale and D.E. Neale

Marketing of engineering services, B. Scanlon

Project evaluation and risk analysis, R.K. Corrie

Management of design offices, P.A. Rutter (Ed.)

Construction contracts, S.H. Wearne

CIVIL ENGINEERING MANAGEMENT

Civil engineering insurance and bonding

Peter Madge, LLM, ACII, FCIArb

T Thomas Telford Ltd, London

Published by Thomas Telford Ltd, Telford House, PO Box 101, 26-34 Old Street, London EC1P 1JH, England

First published 1987

British Library Cataloguing in Publication Data
Madge, Peter
 Civil engineering insurance and bonding – (Civil engineering management)
 1. Insurance, Business—Great Britain
 2. Construction industry—Great Britain
 I. Title   II. Series
    368   HG8059

ISBN 0 7277 0371 4

© Peter Madge, 1987

All rights, including translation, reserved. Except for fair copying, no part of this publication may be reproduced, stored in a retrieval system or transmitted in any form or by any means, electronic, mechanical, photocopying, recording or otherwise, without the prior written permission of the publisher. Requests should be directed to the Publications Manager at the above address.

Set in 11/12pt Plantin by MHL Typesetting Limited
Printed and bound in Great Britain by Billing and Sons Limited, Worcester

Preface

This book is written for engineers to explain some of the basic principles and practices involved in insurance for civil engineering contracts.

It concentrates on four main areas of insurance, namely the insurance of the contract works during construction, public liability, employers' liability and professional indemnity insurance.

Most civil engineering contracts follow a similar pattern. They contain clauses making the contractor responsible for injury or damage in certain circumstances and require him to insure against that injury or damage. Thus there will be clauses stating that the contractor is responsible for loss or damage to the contracts works in certain circumstances and must arrange insurance against that loss or damage. There will be an indemnity clause by which the contractor indemnifies the employer against claims for injury to people or damage to their property. This will be followed by an insurance clause requiring the contractor to take out employers' liability and public liability insurance. Although the words of these clauses differ from contract to contract, the general principles remain the same. In writing this book, therefore, I have taken the ICE Conditions of Contract (5th Edition (June 1983, revised January, 1979)) as a guide. In connection with sub-contractors' responsibilities, I have taken the Federation of Civil Engineering Contractors Form of Sub-Contract (Revised September 1984) as a guide.

Insurance is complicated and construction insurance more than most. Very often the problems are ones of definition. When it comes to interpreting the meaning of engineering contracts or insurance policies, the English language is not always sufficiently precise, and arguments as to the interpretation of words or phrases are bound to arise, particularly where very large sums of money are at stake. Those engaged in the insurance industry would say that the fault is not all of their own making since they have to respond to words and clauses in construction contracts which very often themselves lack precision. Moreover, such clauses have often been drafted without reference to the insurance industry, a common problem with many Standard Form Contracts. Fortunately this situation is being remedied, although slowly. No wonder the engineer regards insurance with some feeling of trepidation.

Constraint
A constraint on this book was the necessity to keep to a maximum of 30,000 words. It is not easy to cover such a wide area in full, given such a brief. Consequently, I have concerned myself in the main with basic principles and practice and have tried to avoid undue technicality in both legal and insurance matters. Nor have I been able to deal with contracts performed outside the United Kingdom, although in the main these do not depart to any great extent from practice in the UK.

Project insurance
I have also approached the book on the basis that, as it is the contractor who in the main carries the responsibilities, it is he who arranges the insurances. There is an alternative way by which the employer arranges the insurances while still leaving the contractor with his contractual responsibilities and liabilities. Often what is known as project insurance is arranged, that is to say, the employer will arrange insurance in the joint names of himself and all contractors. There are many advantages in such an arrangement, particularly for large projects with many contractors involved, but space does not permit any detailed consideration of this type of insurance.

Insurance departments
The insurance market is divided into departments. This is because some insurers or underwriters prefer to specialise in certain risks and have no knowledge of others. Thus insurers specialising in public liability insurance may not underwrite

insurance of the contract works, and insurers specialising in insurance of the contract works may not underwrite professional indemnity risks. This unfortunately leads to the result that it is often necessary to effect several insurances to cover one contract, with the possible danger of gaps appearing in the overall protection. It also explains why it is not possible to have one policy covering 'everything', a question often asked. Trying to obtain complete overall protection is like trying to fit the pieces of a jigsaw together, and since all policies contain limitations or exclusions of some kind the fit is often imperfect.

I have included two chapters on professional indemnity and professional indemnity insurance because of the growing interest in this subject. It is a matter of concern to most professional people, and particularly those employed in the construction industry, that claims for breach of professional duty are on the increase and becoming more expensive to handle and settle. That is why there has been such a sharp increase in professional indemnity premiums and why some sections of the insurance market, having sustained large losses, have discontinued underwriting this class of insurance.

Specimen policies

Some specimen insurance policies are included in the Appendices. These have to be treated with care, since they are meant to be no more than typical insurance policies which will be met in practice. There is no such thing in the insurance world as a standard contractor's policy. Each insurer or underwriter will use his own policies and they will differ. In practice they are often altered to suit the particular requirements of the insured or his professional advisers. Often there may be an element of packaging by which several risks are included in one policy.

Importance of insurance

Insurance plays an important role in all construction projects and many of the things that go wrong are eventually paid for by insurance. Lack of insurance or inadequate insurance can often cause extensive delays and in some cases liquidation of the contractor. It is, therefore, important that the insurance industry be consulted at an early stage for advice and assistance and certainly before the contract conditions are finally drafted. Those insurers, brokers and consultants who specialise in construction insurance have much to offer in the areas of risk identification, analysis and insurance.

Many of the problems that have arisen in construction projects could have been avoided if professional insurance advice had been taken in the early stages. The insurance industry also has much to offer in the field of risk management. It would be wrong to regard the insurance industry as merely the providers of insurance protection – their role is much wider than that.

For more knowledge of the subject, and particularly a more detailed analysis of what insurance policies cover, the reader is referred to the books mentioned in Appendix 7.

The drafts of the book were read by F.N. Eaglestone, P.J. McBrien and J. Powell. I would like to express my thanks and appreciation to all of them for the helpful comments they made and to my secretary Mrs A. Meredith for typing the drafts and dealing so efficiently with my notes and tapes.

Peter Madge
10 Trinity Square
London EC3P 3AX
April 1986

Contents

1 **The insurance market** 1
Some basic insurance principles; insurable interest; disclosure of material facts; contracts of indemnity; subrogation and contribution

2 **The basis of legal liability for personal injury or damage to property** 7
Negligence; nuisance; trespass; breach of statutory duty; strict liability; product liability; liability under contract; the employer's liability for the acts or omissions of his independent contractors; defences; conclusions

3 **Damage to persons and property – an analysis of Clause 22** 19
Employer's indemnity from contractor; purpose of indemnity clauses; risks covered; risks not covered; employer indemnifies contractor for some risks; a confusing clause; the time for suing under Clause 22

4 **Insurance against damage to persons or property – an analysis of Clause 23** 25
Detailed analysis; why insurance is necessary; liability to third parties; does not detract from indemnity; clause cannot be complied with; limits of cover; indemnity to employer

5 **The contractor's public liability policy** 30
Section 1 – the opening clause; section 2 – the insuring clause; section 3 – the limit of indemnity; section 4 – the policy exceptions; section 5 – endorsements; section 6 – the schedule; other amendments that may appear to the policy; sub-contractors; policy conditions

6 **Accident or injury to workmen – an analysis of Clause 24** 42
Overlap with Clause 22; contractors' employees; workmen; no requirement to insure; evidence of insurance

7 **The contractor's employer's liability policy** 44
Who is an employee?; section 1 – opening clause; section 2 – employees; section 3 – the indemnity; section 4 – legal costs; section 5 – the schedule; section 6 – policy conditions; other points of interest

8 **Care of the works – an analysis of Clause 20** 52
Detailed analysis; excepted risks; permanent works; temporary works and works definitions; responsibility for care of works; certificate of completion, use or occupation

9 **Insurance of works – an analysis of Clause 21** 56
Detailed analysis; contractor to insure; joint names; full value; all loss or damage; approval of policy

10 **The cover given by a contractor's 'All Risks' policy** 60
Section 1 – damage to the Works; section 2 – extensions; section 3 – exceptions; some other exceptions; section 4 – limits of liability; section 5 – general conditions; section 6 – claims conditions; section 7 – the schedule

11 **Remedies on contractor's failure to insure – an analysis of Clause 25** 70
Satisfactory evidence of insurance; policies or certified copies; dangers of certificates or other satisfactory evidence; problems of employer insuring if contractor fails to

12 **Sub-contractors and the sub-contract Agreement** 72
Indemnity to contractor; limitations on indemnity; insurances needed; the contract works; sole risk; sub-contractor to arrange own cover; benefit of main contractor's policy

13 **Performance bonds and some of the legal principles governing them** 76
Counter-indemnities; sub-contractors; who issues the bond; underwriting requirements; other forms of bond; bonds are not insurance

CONTENTS

14	**The basis of legal liability for breach of professional duty**	82
	Standard of care; liability to client; reasonable care and skill; liability to third parties; limitation and exclusion clauses; contractor's design and construction contracts	
15	**The cover given by a professional indemnity policy**	86
	Claims made policy; limits and legal costs; extensions to the policy; contractor's design and construction contracts; policy conditions; the interface between various policies; public liability; professional indemnity; contract works policy	
16	**Risk management**	94
	Identification; analysis; treatment and control of risk; damage to property and loss of profit or income, or delay; public liability; health and safety	

Appendices

1.	**Contractor's public liability policy**	97
2.	**Contractor's employer's liability policy**	102
3.	**Contract works policy**	106
4.	**Form of bond**	115
5.	**Professional indemnity insurance for consulting engineers**	118
6.	**Professional indemnity insurance for engineering contractors' design activities**	125
7.	**Further reading**	130

1 The insurance market

The insurance market consists of the insurance companies and the underwriting syndicates at Lloyd's who underwrite the risks; the insurance brokers acting for the insured who place business with the insurers; and the loss adjusters appointed on behalf of the insurers to investigate and negotiate settlement of claims.

There are hundreds of different insurers both in the United Kingdom and outside, in addition to many Lloyd's underwriters, competing for insurance business. They do not compete only on price. The scope of the cover they offer, the service they are prepared to give and their reputation also comes into play. The insurance industry in the late 1970s and early 1980s went through very difficult times, incurring large underwriting losses. The financial stability of the chosen insurer is thus an important point. The insurers must be in existence at the time of the claim.

Insurers in the market
Not all insurers have any deep knowledge of the construction industry. Many who do not may simply be attracted by the premiums involved. Insurance companies are run, like any other business, with a view overall to profit. Insurers have to have an eye on the market place, knowing when to come in to a particular line of business, when to stay in, when to increase the premiums and when to come out. Therefore, some insurers may be expanding and willing to underwrite new business, while at the same time other insurers may be pulling out of a market in which they have lost money. Some insurers may be underwriting for underwriting profit, that is, a surplus of premium over claims. Other insurers may be underwriting purely for investment income. There is a time lag between the date of payment of the premium

and the settlement of the claim, and this money expertly invested can produce income, particularly when interest rates are high.

The insurance broker is the intermediary between the insured and the insurer and acts as the insurer's professional adviser. His role is to find the most competitive insurers prepared to underwrite the risk, compatible with wide cover and financial stability. His duty is to the insured. The insurance broker who is an expert in construction business will know the various policies on offer by the various insurers and – what is more to the point – how to alter them to reflect his or his client's individual wishes.

The loss adjuster is an independent person appointed by the insurer to investigate, negotiate and settle claims under the insurance policies.

Every year billions of pounds change hands as insurers receive premiums and pay out claims. Many of the risks underwritten in the insurance market are enormous. At the time of writing this book discussions on constructing a Channel tunnel linking England with France are under way. The amounts at risk and the liabilities involved are so huge they will stretch the worldwide insurance market to its full capacity.

High financial risk

In many engineering contracts the value of the contract works is high. Moreover, in view of the litigious times in which we live, claims for common law damages for people who have sustained injury or damage to their property are on the increase. In 1985 a record amount for the UK of £679,000 was awarded for personal injuries, which shows how high the figures can reach. These risks are invariably covered by insurance, and the payment of very high amounts of money may often depend upon the interpretaton given to a word or phrase in such a policy, or the information given by the insured when proposing the insurance to the insurer or completing his proposal form. Unfortunately many businesses have a casual approach to insurance. It has a low order of priority. Often insufficient thought is given to the matter by the insured, and the information given to the insurer to underwrite the risk, and the replies to questions asked by the insurer are often incomplete or inadequate. This can have a disastrous effect on the validity of the policies when the claims come to be made, as will be explained later.

Some basic insurance principles
The subject matter of an insurance policy may be the property owned by the insured against which he wishes to insure, i.e. the contract works, or the creation of a legal liability against him for which he wants insurance protection, i.e. public liability.

Insurable interest
The basic principle of insurance law is that before a person can effect an insurance policy he must have an insurable interest in the subject matter. By that is meant that he must suffer some detriment if the subject matter is damaged, or must benefit by its well-being. Thus the owner of property is prejudiced if it is lost or stolen. Person A cannot insure person B's property against damage, since in the event of it being lost or stolen, person A has no interest in it, and it is a matter of no consequence to him. If person A were able to insure person B's property, person A would clearly benefit by its destruction. Any of insurance monies payable to person A would be contrary to public policy and such a policy would thus be illegal.

Like other forms of contract, insurance is subject to the normal contractual rules of offer and acceptance, consideration (the premium), legality, agreement of the parties, contractual capacity of the parties and the intention to create a legal relationship. Insurance contracts, however, differ fundamentally from other commercial contracts in the sense that they are bound by the principle of utmost good faith. There is a duty on the insured to disclose to the insurer all material facts bearing on the risk.

Disclosure of material facts
A material fact is something which would influence the judgement of a prudent insurer in agreeing to accept the risk or not and in deciding the amount of premium he would charge. The test of whether a fact is material or not is whether it would have influenced the judgement of a prudent insurer, not the particular insurer issuing the policy. Whether or not the insured considered the fact to be material is not relevant. It is the test of the prudent insurer, not the prudent insured.

The insured has a duty to disclose all material facts which have a bearing on the risk proposed and must make no misrepresentation about those facts or the risk. The onus of proving that there has been a non-disclosure of information or fact is upon the insurers. The insured must disclose all those material facts which are within his knowledge, whether actual or presumed. He must disclose all those facts which he knows, or ought in the ordinary course of business affairs to know or have known about.

This represents the common law position. Insurers often insist upon a proposal form being completed before insurance is offered. This asks a number of relevant questions which are material to the risk but the important point to note is that even though a question is not asked, if there is something material to the risk then the insured must disclose it. Moreover, most proposal forms contain a declaration and warranty at the foot of the form saying the insured has answered all questions accurately and not withheld material information. The insured has to sign this. The effect of signing this declaration and warranty is that the insured warrants the accuracy of all the answers on the proposal form and further warrants that he has not withheld or failed to disclose all material facts. A warranty in insurance law has a strict interpretation. Thus, any inaccuracy on the proposal form or non-disclosure of material information gives the insurers the right, if they so wish, to treat the policy as void. Hence the importance of making sure that all answers to questions and information shown on proposal forms are correct.

- Over the years many technical and legal rules have been established in relation to non-disclosure. Generally the following will be held to be material and must, therefore, be disclosed:
  - facts indicating that the subject matter of the insurance is exposed to more than the ordinary degree of risk, for example that the works are being constructed in an area susceptible to flooding;
  - facts indicating that the insured is activated by some special motive as, for example, where he greatly over-insures the value of his property;
  - facts showing that the liability of the insurer is greater than he would normally have expected it to be as, for example, that very valuable property is contained in a building with a poor standard of security and in an area where there has been a high incidence of theft;

INSURANCE MARKET

- o facts showing that there is a moral hazard attaching to the insured suggesting that he is not a fit person to whom insurance can be granted, for example a person with a bad criminal history involving theft or arson; and
- o facts which to the insured's knowledge are regarded by the insurers as material, such as a high incidence of theft on a contract site.

• On the other hand there are facts which, although material, may become immaterial in certain circumstances and there is, therefore, no obligation upon the insured to disclose them. For example:

- o facts which are already known to the insurers or which they may be reasonably presumed to know, for example that work at height or depth is hazardous;
- o facts which the insurers could have discovered themselves by making some enquiries; for example, if the insurers are told that demolition work is to be carried out, they have the opportunity to investigate the risks involved if they so require;
- o facts where the insurer has waived further information, where for example he is told that a certain contractor is being used to perform the work and that contractor is known to be unsatisfactory;
- o facts tending to lessen the risk – for obvious reasons, since anything that lessens the risk is beneficial to the underwriter.

Contracts of indemnity

Contract insurances are contracts of indemnity under which the insurers agree to make good the insured's loss by payment, repair, reinstatement or some other method. The principle in theory states that the insured should neither be better off nor worse off after the occurrence of an insured event but – in so far as a sum of money can do so – should be placed in exactly the same position as if the event insured against had not happened. If property worth £500 is lost then the insurers give the insured £500. He is thus indemnified. If it is insured for £900 he still gets £500. To give him more than his loss is to encourage him to bring about the insured event. The onus of seeing that the sum insured is correct is on the insured.

Subrogation and contribution

The matter of indemnity is subject to two important principles known as subrogation and contribution. Subrogation means that where an insured has been indemnified, then the insurers can stand in his shoes and avail themselves of all rights and remedies open to the insured. In other words if the insured's property worth £500 has been damaged by negligence of a third party, the insurers will pay the insured £500. The insured will have been indemnified. They are then entitled in the insured's name to sue the third party to recover the £500.

Contribution applies where the same interest and the same risk is insured by more than one policy. In that case both policies contribute towards the loss. The insured cannot recover his loss in full under both policies, as that would defeat the principle of indemnity.

2 The basis of legal liability for personal injury or damage to property

Accidents occur frequently on construction projects. People are injured or their property is damaged. These people, however, have no automatic right to obtain damages from those who caused the injury or damage. They must first prove legal 'fault'. The basis of this legal 'fault' is explained below. Once the basis of legal liability has been established against the defendant, he will expect to be protected by his liability insurance policies if they have been prepared correctly.

How liability arises

- There are several ways in which civil liability may attach to the defendant and it is for the plaintiff to bring his claim under one or more of these headings. These are:
  - negligence
  - nuisance
  - trespass
  - breach of statutory duty
  - strict liability
  - vicarious liability for employees or contractors
  - liability under contract.

Other than the last item, liability under contract, these forms of liability are often referred to as *torts*, i.e. civil wrongs independent of contract. It is necessary to have some understanding of the legal issues involved to understand how liability arises, how it may be excluded, modified, apportioned or transferred under appropriate clauses in the building or engineering contract and the extent of cover available under legal liability insurance policies.

Negligence
Negligence produces most claims.

Negligence is omitting to do something which a reasonable and prudent man would do or doing something which a prudent and reasonable man would not do. Negligence on its own, however, will not bring liability. A contractor who is the sole occupant of an island can be as negligent as he wishes since there is no other person at risk. It is only when other persons or property are at risk that the contractor will owe them a duty of care not to be negligent.

- To succeed in an action for negligence the plaintiff must prove:
  o that the defendant owed him a duty of care;
  o that the defendant broke that duty of care; and
  o that the plaintiff suffered injury or damage as a result of the breach.

Duty of care
The defendant owes a duty of care to those whom he ought reasonably to have in mind at the time he is carrying out his activities. This is the 'neighbour' principle, established in the case of *Donoghue* v. *Stephenson* (1932), where Lord Atkin answering the question 'Who in law is my neighbour?' said, 'Persons who are so closely and directly affected by my act that I ought reasonably to have them in contemplation as being so affected when I am directing my mind to the acts or omissions which are called in question'.

Whether there is a duty of care is a question of law to be decided by the court at the time.

Reasonable care
The extent of the duty of care depends upon reasonable foresight and reasonable care. The defendant will owe a duty of care to those persons whom he can reasonably foresee may be injured by his acts or omissions but the defendant will only break his duty of care if, reasonably foreseeing the risk of loss or damage to his neighbour, he does not take reasonable care to avoid it.

The standard of reasonable care is an objective one, being that of the ordinary reasonable man in the circumstances of the case.

The burden of proving negligence rests upon the plaintiff except where the principle of *res ipsa loquiter* (the thing speaks for itself) applies. If a sack of flour falls out of a building on to an innocent bystander *res ipsa loquiter* applies. In the natural order of things, sacks of flour ought not to fall out of buildings. Negligence is, therefore, presumed against the defendant and the onus of disproving it passes to the defendant. The defendant may rebut the presumption by showing that the accident could reasonably have occurred without any negligence on his part.

Even though a person may be negligent, the plaintiff will not succeed if his injury or damage was too remote. There has to be a causal link between the act of negligence and the injury, loss or damage suffered that is not too remote in the eyes of the defendant. A motorist who injures a child on holiday in London will not be liable for the shock to the mother at home in Glasgow when she is told the news several days later. From the defendant's point of view the injury is too remote: it is too far removed from the act of negligence to create liability.

Nuisance

Nuisance is the wrong done to a person by creating noise, smells, dust or vibrations, thus unlawfully disturbing him in the enjoyment of his property or, in some cases, in the exercise of a right. The basic principle of the law relating to nuisance is that a person must so use his property so that he does not cause harm to others.

The basic principle of nuisance is that the plaintiff is adversely affected in his use or enjoyment of *land* or some *right* over it.

Nuisance may be public or private.

A *public* nuisance is a crime. A nuisance will not be a public one unless the number of persons affected is substantially large enough to constitute a 'class' of persons, e.g. a large number of people affected from blasting and quarrying operations near their home. A public nuisance is only actionable as a civil claim if the plaintiff can show that he has suffered *particular* damage over and above that suffered by the rest of the public at large.

A *private* nuisance is an interference for a substantial length of time by owners or occupiers of property with the use and enjoyment by others of neighbouring property. A private nuisance can consist of allowing the escape of noxious things such as smoke, smells, noise, gas, vibration, damp and so on, so as to interfere with the health, comfort or convenience of neighbours in the enjoyment of their property, or the wrongful disturbance of rights attaching to land, such as a right to light and air, rights to support of land and buildings and rights in respect of water and rivers. Liability for such disturbances of rights attaching to land is strict.

Nuisance and negligence may overlap. Although some actions in nuisance arise from intentional acts, some also arise from the defendant's inadvertence and in these cases the plaintiff may sue in nuisance or negligence.

Trespass

Trespass is an unlawful act committed with force and violence, on the person, property or right of another.

The essence of trespass is the *direct* injury. Indirect injuries fall within the areas of negligence or nuisance. Trespass is a strict tort and is actionable *per se*, that is without proof of special damage. It is trespass for a contractor to go on to another person's land without lawful authority, even if he causes no damage.

Trespasses may be of three kinds – to *land*, to *goods* and to the *person*.

Trespass to land

An unlawful entry on to land or buildings in the possession of another is trespass. If a person remains on the land or allows things he has unlawfully placed on the land to remain there, such trespass can be the subject of a fresh action for each day that it continues. Trespass in such circumstances is a continuing trespass. A possessor of land possesses the soil beneath and the column of air above it. An occupier may remove things trespassing into his air space even though they do no harm, e.g. overhanging branches or tower cranes.

Trespass to goods

Trespass to goods consists of a direct and unlawful injury to or interference with goods in the possession of another.

The point about possession is essential because it is the sole basis of the plaintiff's right. Only the person in possession of the goods at the time of the trespass can bring the action.

BASIS OF LEGAL LIABILITY

Trespass to person
Trespass to a person consists of assault or battery or false imprisonment. Battery is the application of unlawful force to the person of another. Assault is putting a person in reasonable fear of immediate battery.

Breach of statutory duty
Acts of Parliament often contain regulations designed to prevent injury to persons or damage to property, or to give to persons the legal right to sue for a breach of the regulations contained in the legislation. Examples are requirements of the Factories Acts, the Construction Regulations, the Health & Safety at Work Act, the Building Regulations, the Occupier's Liability Act, the Fatal Accidents Act and the Mines & Quarries Acts. Although such legislation may lay down criminal sanctions e.g. fines, a breach of any duty by the defendant under such legislation may be actionable in civil law by the person who is injured, but the success of his claim will depend on a construction of the words of the statute. Thus if there is a duty to fence dangerous machinery, a failure to fence will be a breach of the duty and the breach may be absolute. Alternatively, legislation may require fencing 'insofar as it is reasonable or practical to do so'. This is not absolute. An enquiry has to be made into whether the fencing at the material time was reasonable or practicable. In other words, the risk has to be balanced against the practicability of the measures needed to avoid it. Cost, for example, has to be taken into account.

- The *burden of proof* that there has been a breach of statutory duty is upon the plaintiff. The standard of proof required in civil actions is that on a balance of probabilities the defendant has been in breach. In addition, the plaintiff must prove:
    o that the statute was broken. For example, in one case the plaintiff failed to prove that his skin disease arose from his employment in sandpapering wood.
    o that the breach of the duty caused his injury, in other words that there was a causal link. For example, a steel erector fell and was killed because he was not wearing a safety belt in accordance with statutory regulations. It was the practice of steel erectors never to wear belts, thus if belts had been available it was improbable that the deceased would have

11

worn one. The claim failed, as the breach of duty in no way caused the accident.
- that the plaintiff was one of a class of persons the statute was intended to protect. For example, an engine driver was injured when a car ran into a level crossing gate which had not been maintained in accordance with the statutory requirements. His claim failed because the purpose of the Act was to protect *road* users, not engine drivers.
- that the injury of which the plaintiff complains is one the statute was intended to prevent. For example, a plaintiff's sheep were washed overboard because the defendant had failed to provide pens on the ship's deck as required by statute. The claim failed. The object of the statute was to prevent the spread of disease amongst sheep, not to prevent them from being washed overboard.

Negligence and breach of statutory duty are usually regarded as two distinct claims even though there may be some overlapping – for example, where the statutory duty is to 'take all such circumstances as in the case are reasonable'. In negligence the standard of care is decided by the court, whereas in breach of statutory duty the standard of care is actually *laid down in the statute*.

Strict liability
A defendant may have liabilities imposed upon him by law which he cannot avoid even though he has exercised all reasonable care and skill. The liability is said therefore to be strict, since it does not require proof of negligence or lack of care or wrongful intention. Strict liability can occur in the circumstances such as those covered by the rule in *Rylands* v. *Fletcher* (liability for escape of dangerous things on defendant's land). In this case, the employer engaged an independent contractor to construct a reservoir on the employer's land. Unknown to the employer there were disused mine shafts on the site and when the reservoir was filled the plaintiff's adjoining coal mine was flooded. The employer was held liable since the court held that 'any person who for his own purposes brings on his land and collects and keeps there anything likely to do mischief if it escapes must keep it at his peril and if he does not do so is *prima facie* answerable for all the damage which is the natural consequences of its escape'.

BASIS OF LEGAL LIABILITY

Whether a 'thing' which has been brought on to the defendant's land is one which is likely to do mischief if it escapes is a question of fact. A wide variety of things have been held to be within the *Rylands* v. *Fletcher* rule, such as electricity, gas, sewage, vibration, poisonous trees, gipsies and caravan dwellers.

For the rule to apply there must be an *escape* from the defendant's land of the 'thing likely to do mischief' and there must have been a non-natural use of the defendant's land. In one case the defendant had explosives on his land. The plaintiff, a munitions inspector of the Ministry of Supply, was injured, while on the premises, by an explosion. He had no claim under the *Rylands* v. *Fletcher* rule since his injuries had taken place on the land. There had been no *escape* from the land of the dangerous thing. Also there must be a non-natural use of the land, so things naturally on the land, such as weeds, are not within the rule.

From the employer's point of view, the principle of strict liability can be seen from *H&N Emanuel Ltd* v. *Greater London Council & Another* (1971) where the Council engaged contractors to demolish some prefabricated buildings. Contrary to express instructions from the Council, the demolition contractors burned some of the debris in a bonfire and sparks set fire to surrounding buildings. The GLC were held liable for the negligence of their contractors.

Animals

The Animals Act 1971 codified the previous common law relating to animals.

There is a strict liability for injury caused by dangerous animals (defined as fully grown animals which, unless restrained, are likely to cause severe damage). The liability is imposed on the 'keeper', that is, the owner or the person who has the animal in his possession.

Guard dogs on contract sites could come within the rule. Trespass by cattle may also be strict.

Fire

Strict liability may attach to certain forms of damage arising out of fire, and liability for fire is often governed by the rule in *Rylands* v. *Fletcher* or by statute.

An employer is liable for fire damage caused by the negligence of his independent contractors. In *Balfour* v. *Barty-King* (1956), the occupier of a house engaged contractors to thaw her water pipes. They used blow lamps and set fire to the house. The fire spread to the neighbouring house causing damage. The occupier was held liable.

Product liability
An EEC Directive on Product Liability will become law by July 1988. It will introduce strict liability on producers of goods which are defective, causing personal injury or damage to personal property.*

Liability under contract
Liability under the torts considered previously is a liability *imposed upon the parties by law*. Under the terms and conditions of written contracts, liabilities are imposed upon or accepted by the parties voluntarily by the terms of that contract. Thus parties may add to or detract from their tort liability. Such contracts often contain indemnity clauses of the type seen in the ICE Conditions where the liability of the parties may be widened.

The employer's liability for the acts or omissions of his independent contractors
There is a distinction in law between an employee who is a person employed under a contract *of* service and an independent contractor who is a person employed under a contract *for* services.

An independent contractor is a person engaged by the employer to produce a given result where the employer has no power to direct the method of doing the work. This is left to the discretion of the independent contractor because he has the necessary skills. On the other hand, an employee is usually subject to the control of his employer and must do what he is told.

An employer is normally vicariously liable for the activities of his employees because he has the power of control over them and thus will be liable for their acts or omissions during the course of their employment. He is not, as a matter of general principle,

---

* In December 1986 the Consumer Protection Bill was in the House of Lords introducing Strict Liability.

BASIS OF LEGAL LIABILITY

liable for the activities of his independent contractors since normally the control factor is missing. There are, however, important exceptions to this principle and the employer may be liable for the acts of his independent contractors provided that the liability arises from the work itself. In these circumstances the employer is under a strict duty to avoid injury or damage. Examples of these types of circumstances are given in the following six subsections.

Where the work involved 'extra hazardous' operations
The defendant was employed by the plaintiffs to take flashlight photographs in a third party cinema. The defendant negligently caused fire for which the plaintiffs had to compensate the cinema owner and the court held that the plaintiffs could recover this amount from the defendants because the plaintiffs were liable to the cinema owners for their independent contractor's extra hazardous act.

Work in connection with the withdrawal of support to land or buildings would come under the 'extra hazardous work' definition. So too would work which is being performed on or close to a highway, for example a demolition contract.

Where the contractor is employed to do illegal or unlawful work
The defendants employed a contractor to dig trenches in the streets, without obtaining the necessary authority. The plaintiff fell over stones left by the contractors and the defendant was held liable.

Where a measure of control over the contractor is retained by the employer through the provision of machinery or men
Sub-contractors were used to discharge ships. The employers provided equipment to enable them to do this. People engaged by the sub-contractors were injured. The employer was held liable because he had retained a sufficient degree of control over the workmen by the provision of the equipment.

Where the employer is under a statutory duty to perform the work in a particular way
By statute the defendants were authorised to construct a bridge across a river but subject to the requirement that the bridge should not detain any vessel using the river longer than was

necessary. The bridge could not be opened for several days because the independent contractors did not do the job properly. The employers were held liable.

Where the negligence is that of the employer in using an inexperienced or incompetent contractor or in failing to give sufficient instructions
The employer was held responsible for the activities of contractors who had deposited sewage from a cleaning contract on to another person's land. The employer should have taken more care to see that the contractor's disposed of the sewage correctly.

Where there is liability for fire
The defendant employed contractors to thaw some pipes with a blow lamp. They negligently set fire to the plaintiff's premises. The defendants were liable for their contractor's negligence.

For the employer to be liable the tortious activities of the independent contractor must be ones upon which he is employed to work and not merely ancillary acts connected with it. In other words, the employer is not liable for collateral negligence of the independent contractor. In one case the defendants were main contractors. Their sub-contractor's workmen injured the plaintiff by carelessly leaving a tool on a window sill from which it fell onto the plaintiff. The court held that the main contractors could not be held liable for the simple collateral negligence of the sub-contractor's workmen.

Defences
Several defences to the claim may be open to the defendant, as follows.

Denial
The obvious one is that of denial whereby the defendant denies that he was negligent or in breach of any duty owed to the plaintiff.

Inevitable accident
If an accident has taken place despite the exercise of all reasonable care on the part of the defendant this defence may arise, although it is not a defence to all torts.

In negligence cases this is an argument of limited application. If the defence is that the accident was inevitable, it is another way of saying that the defendant has not failed in any duty to observe the standards of care expected of him.

Act of God
This is a defence of limited application being applicable to the torts involving strict liability such as the rule in *Rylands* v. *Fletcher*. An Act of God is an event that is due to natural consequences directly and exclusively without human intervention and that could not have been prevented by any amount of foresight and care. (Incidentally the view that insurance policies do not cover Acts of God is wrong. They do.)

Emergency action
The defendant may plead that he acted in an emergency and that his actions were therefore reasonable in all the circumstances as, for example, where a motorist swerves to avoid a young child running into the road.

Contributory negligence
Where the plaintiff is partly to blame for his injury or damage his damages will be scaled down accordingly e.g. plaintiff 30% to blame; damages reduced by 30%. This is not really a defence unless the amount of contributory negligence on the plaintiff is 100%, in which case it is clear there was no fault on the defendant.

Volenti non fit injuria
A literal translation is 'to him who is willing there can be no injury'. In other words, if the plaintiff consented to undergo the risk he cannot later sue if as a consequence of running the risk he sustained injury or damage. The consent, however, must be real. A person cannot be taken to have consented to a risk the nature of which he did not really understand, e.g. a boxer accepts the risk of injury but not if his opponent has weights in his gloves.

Limitation of action
There are time limits for bringing claims prescribed by legislation. If the claim is not brought within the appropriate time limit, then the defendant may argue that the claim is time-barred.

Statutory authority
Where a statute specifically authorises a certain act to be done, injury or damage arising out of the performance of that act may carry with it statutory exemption from liability. In the early days of steam locomotives the railway companies had an exemption from liability for fires caused by escaping sparks. It is a question of construction as to whether the appropriate statute gives total exemption or partial exemption, i.e. so as not to cover the negligent performance of the statutory function.

Act of a stranger
The true cause of injury of damage may not be that of the defendant's but of a third party over whom he has no control. Thus in one case a third party discharged water into the defendant's reservoir so that it overflowed and damaged the plaintiff's property. The court held that the defendant's were not liable under the principle of *Rylands* v. *Fletcher*.

Conclusions
This brief review of the basis of liability is very much an oversimplification but shows that there is a complex system of legal rules involved in deciding liability and the extent to which damages will be paid.

The extent to which contracting parties may agree between themselves how such liability will be handled and apportioned by appropriate clauses in the contract is considered in the next chapter. The fact that a person may be insured does not affect the liability. Insurance cannot change the law, merely provide money.

Liability insurance policies protect the insured against his legal liability. The extent (if any) of the insured's legal liability has to be investigated and determined. If there is no legal liability, then there is nothing against which the insurers will have to indemnify the insured. If there is, the insurer will pay the claim on behalf of the insured.

3 Damage to persons and property – an analysis of Clause 22

It is an unfortunate fact of life that on many engineering projects people are injured or their property is damaged. Inevitably they sue. There are many potential defendants ranging from the employer, main contractor, sub-contractor, the engineer and other professional consultants. There are many ways in which liability may attach, ranging from negligence, nuisance and trespass to strict liability, liability under contract and liability under statute as was explained in Chapter 2.

Indemnity
As between the contractor and the employer, Clause 22 attempts to determine and apportion liability and its financial consequences. It is an indemnity clause.

Indemnity clauses serve a useful purpose. In the normal way an injured person will sue the person whom he considers to be responsible. Often he will sue more than one. One person sued may bring in another as a joint defendant. Since each to some extent will tend to put some of the blame upon the other, the assessment of liability and its apportionment can be time-consuming and costly. If the normal processes of law were left to take their course the matter could go to litigation, leaving the courts finally to determine who is responsible. This is time-consuming and expensive. Well-drafted indemnity clauses, therefore, serve a useful purpose in that they enable the contracting parties to agree beforehand who will be responsible for what and in what circumstances. They know where they stand and can arrange their insurances accordingly.

CIVIL ENGINEERING INSURANCE AND BONDING

The following is the wording of Clause 22.

'22. (1) The Contractor shall (except if and so far as the Contract otherwise provides) indemnify and keep indemnified the Employer against all losses and claims for injuries or damage to any person or property whatsoever (other than the Works for which insurance is required under Clause 21 but including surface or other damage to land being the Site suffered by any persons in beneficial occupation of such land) which may arise out of or in consequence of the construction and maintenance of the Works and against all claims demands proceedings damages costs charges and expenses whatsoever in respect thereof or in relation thereto. Provided always that:

(a) the Contractor's liability to indemnify the Employer as aforesaid shall be reduced proportionately to the extent that the act or neglect of the Employer his servants or agents may have contributed to the said loss injury or damage;

(b) nothing herein contained shall be deemed to render the Contractor liable for or in respect of or to indemnify the Employer against any compensation or damages for or with respect to:

(i) damage to crops being on the Site (save in so far as possession has not been given to the Contractor);

(ii) the use or occupation of land (which has been provided by the Employer) by the Works or any part thereof or for the purpose of constructing completing and maintaining the Works (including consequent losses of crops) or interference whether temporary or permanent with any right of way light air or water or other easement or quasi easement which are the unavoidable result of the construction of the Works in accordance with the Contract;

(iii) the right of the Employer to construct the Works or any part thereof on over under in or through any land;

(iv) damage which is the unavoidable result of the construction of the Works in accordance with the Contract;

(v) injuries or damage to persons or property resulting from any act or neglect or breach of statutory duty done or committed by the Engineer or the Employer his agents servants or other contractors (not being employed by the Contractor) or for or in respect of any claims demands proceedings damages costs charges and expenses in respect thereof or in relation thereto.

(2) The Employer will save harmless and indemnify the Contractor from and against all claims demands proceedings damages costs charges and expenses in respect of the matters referred to in the proviso to sub-clause (1) of this Clause. Provided always that the Employer's liability to indemnify the Contractor under paragraph (v)

DAMAGE TO PERSONS AND PROPERTY

of proviso (b) to sub-clause (1) of this Clause shall be reduced proportionately to the extent that the act or neglect of the Contractor or his sub-contractors servants or agents may have contributed to the said injury or damage.'

The words in brackets in the first line (except if and so far as the contract otherwise provides) emphasise that the contract should be read as a whole, since words elsewhere in the contract may qualify the indemnity.

A confusing clause

Unfortunately Clause 22 can lay no claim to being a well-drafted clause. It is in fact the reverse. It is confusing and contradictory in places and is well in need of being redrafted as will be seen later. Indeed it is difficult to see how in a small book of this nature the clause can be truly explained without becoming extremely legalistic. What follows, therefore, is an attempt to give a brief overview of the clause using the minimum of legal technicalities.

The effect of Clause 22(1) and the first sub-paragraph (a) are reasonably clear. The contractor gives to the employer a very wide form of indemnity against all forms of injury or damage to any person or property and against all legal proceedings, costs and charges. However, under proviso (a) the indemnity is reduced to the extent that the employer, his servants or agents may have contributed to the injury or damage. Thus in a claim against the employer where the contractor is 80% at fault and the employer is 20% at fault, the contractor is responsible for only 80% of the claim.

It is made clear that the contractor's indemnity does not apply to the works (for which responsibility for loss, damage and insurance under a joint names policy effected by the contractor are specified in other clauses).

- The indemnity does not apply to:
  - 'damage to crops on the Site' which will, therefore, be the employer's responsibility.
  - 'the use or occupation of land ... or interference ... with any right of way light air or water or other easement ... which are the unavoidable result of the construction of the Works'. This is a logical requirement since if such forms of

injury or damage are unavoidable there is no reason why the contractor should be responsible for them. There is also the additional point that the contractor cannot insure his liability against unavoidable injury or damage, since his public liability policy will be restricted to *accidental* injury or damage, a point explained later in Chapter 4. An easement is the right of one person to use another person's land. Where two pieces of land, but not the land in between, are owned by one person and he uses a path to go from one to the other, the owner is using this right in his capacity as the owner of both pieces of land and this right over the land he does not own is called a quasi easement.

o 'the right of the Employer to construct the Works ... on over under in or through any land'. This again seems a logical exemption since it is the employer's responsibility to make sure that he has good title to the land upon which the contract is to be performed and has adequately checked all deeds, leases and planning permissions.

o 'damage which is the unavoidable result of the construction of the Works in accordance with the Contract' is again another sensible exemption. If such damage is unavoidable then it is a risk that the employer should bear. Note that the reference is to damage, not personal injuries.

o 'injuries or damage ... resulting from any act or neglect ... by the Engineer or the Employer his agents or servants ...'. The confusion here is that while in the opening paragraph of Clause 22(1) the contractor's indemnity to the employer is *reduced* to the extent that the act or neglect of the employer has contributed to the injury or damage, we now have a statement that the contractor shall not be liable *at all* where the act or neglect of the employer has contributed. Also the words 'breach of statutory duty' and reference to 'the Engineer' now appear, whereas they did not in Clause 22(1)(a).

A further confusion in Clause 22(1) is that the contractor indemnifies the employer against 'all losses and claims for injuries ... to *any* person ...' which must include the contractor's employees. Yet Clause 24 deals with the contractor's employees. There is thus an overlap. Presumably Clause 24 will apply as it is the more specific clause.

DAMAGE TO PERSONS AND PROPERTY

A brief discussion of the employer's indemnity to the contractor and the question of the time for suing under Clause 22 now follows.

Employer's indemnity to contractor
Under Clause 22(2) the employer indemnifies the contractor against the first four risks mentioned above, presumably because they all deal with forms of inevitable or unavoidable damage against which the contractor cannot obtain insurance and ought not in equity to be liable for anyhow. In so far as the fifth item is concerned, i.e. injury or damage to persons or property, the employer indemnifies the contractor but only proportionately. In other words if the employer is 60% at fault and the contractor is 40% at fault, the indemnity from the employer will cover only 60% of the contractor's liability for injury or damage.

It is a pity that a clause dealing with such important matters is so badly drafted since the sums of money involved can be substantial. In the United Kingdom personal injury awards have now exceeded £600,000 and property damage claims can be substantially more than this, often measured in millions of pounds.

Where two or more persons may be liable for injury, loss or damage, the Civil Liability (Contribution) Act, 1978 provides that any person liable in respect of any damage may recover contribution from any other person liable in respect of the same damage and seek from the court a contribution from others which in the court's view would be 'just and equitable'. Reading Clause 22 as a whole it seems that the apportionment of liability as attempted to be defined therein does not materially differ from the result which would apply under the Civil Liability (Contribution) Act.

The time for suing under Clause 22
The right of one party to claim indemnity from another arises when he has sustained a claim or loss because until then there is nothing against which he can be indemnified. Thus a third party injured during the performance of the contract may sue the employer and obtain compensation many years after the contract has been finished. The employer's right to claim indemnity from the contractor under Clause 22 arises at that time. Public liability insurance policies cover liability for injury or damage *occurring*

during the period of insurance, thus provided the injury which occurred for example in 1980 was reported to the insurers in accordance with the policy conditions, the insurers will indemnify the contractor against the indemnity he has given even though the policy was not in force at the time the claim was made and settled in 1986.

4 Insurance against damage to persons or property – an analysis of Clause 23

Both the contractor and the employer may face claims from third parties alleging injury to their person or property. Clause 22 attempts to decide as between the contractor and employer who will be responsible for such claims. There is little point in deciding who is responsible if that person lacks the financial resources to deal with such claims. Hence the requirement to effect insurance is sensible and that is what Clause 23 does. The clause requires the contractor to take out public liability insurance protecting both himself and to a limited extent the employer. The employer does not get complete protection against all his liability – he is entitled to only the same protection as the contractor is under the policy. If there are shortfalls in the policy the employer is uninsured – see Chapter 5.

The following is the wording of Clause 23.

'23. (1) Throughout the execution of the Works the Contractor (but without limiting his obligations and responsibilities under Clause 22) shall insure against any damage loss or injury which may occur to any property or to any person by or arising out of the execution of the Works or in the carrying out of the Contract otherwise than due to the matters referred to in proviso (b) to Clause 22(1).

Amount and Terms of Insurance
(2) Such insurance shall be effected with an insurer and in terms approved by the Employer (which approval shall not be unreasonably withheld) and for at least the amount stated in the Appendix to the Form of Tender. The terms shall include a provision whereby in the event of any claim in respect of which the Contractor would be entitled to receive indemnity under the policy being brought or made against the Employer the insurer will indemnify the Employer against such claims and any costs charges and expenses in respect thereof. The

Contractor shall whenever required produce to the Employer the policy or policies of insurance and the receipts for payment of the current premiums.'

Detailed analysis
The obligation on the contractor is to effect public liability insurance 'throughout the execution of the Works' i.e. against liability to third parties but not to the Works. If the contractor were to do this it would give neither him nor the employer complete protection. Public liability policies, as will be explained in the next chapter, protect the insured against his liability for injury or damage occurring during the period of insurance. This injury or damage may occur some considerable time after the act or omission giving rise to it. Thus during the execution of the work the contractor may make a mistake e.g. bad wiring or faulty connection, which results in a fire or explosion many years later, causing injury or damage. The claims against the contractor and the employer may not, therefore, be made until some considerable time after the contract has been completed and the insurance has ceased.

Ideally what is required is insurance which operates for an unspecified period of time in the future, but such insurance is not available. If the contractor has an annual public liability policy which is renewed every year (as most do) then for as long as the policy remains in force the contractor and the employer (if the policy is correctly worded) will be protected. If, however, the contractor has taken out an individual policy for the contract (as sometimes happens) then it would be sensible for the contractor, subject to insurer's agreement, to maintain that policy in force for several years after the contract has been completed. For how long it is impossible to say. Claims for injury to the person or damage to property are governed in law by limitation periods. A person has six years to sue, starting from the date his property was damaged.* It is three years for personal injuries but with a wide discretion to the courts to extend it. Nobody can calculate the date at which the damage will take place. As a practical measure, however, it would be sensible for the contractor to maintain his

---

* The Latent Damage Act 1986 allows three years from the date of discovery of latent defects subject to a long stop period of 15 years from the defendant's breach of duty.

INSURANCE AGAINST DAMAGE TO PERSONS

single policy in force for a period of say six years after the end of the contract. The insurer will have to be told that the object of such an extension is to cover the possibility of injury or damage occurring after completion of the contract and for which, therefore, the premium should not be high.

Many contractors who have arranged single project policies cater for this liability by extending their annual policies to cover the 'run-off' liability arising from single policies.

- The obligation to arrange the insurance does not limit the contractor's responsibillities under Clause 22 where the contractor indemnifies the employer against certain forms of injury, loss or damage. The contractor remains responsible under Clause 22 whether or not he has effected insurance.

The insurance need not cover the matters referred to in proviso (b) to Clause 22(1).

- The obligation is to insure against *any* damage loss or injury ... to *any* property ... or to *any* person .... This is a requirement with which the contractor cannot comply. The public liability policy will not cover damage to any person since it will exclude injury to persons under a contract of service or apprenticeship with the contractor (for which his employer's liability policy will apply). Moreover a public liability policy (as will be explained in the next chapter) does not cover any form of damage to any property. The policy has limitations and exclusions and will thus exclude injury or damage arising from radioactivity from any nuclear fuel (uninsurable in the commercial insurance market), injury arising from motor vehicles used on the road (for which compulsory motor insurance is required) and liability arising out of aircraft, ships or vessels. Seldom do the policies issued by insurers correspond exactly with the requirements of Clause 23. A detailed analysis of the policy is, therefore, necessary by both the contractor and employer to see what the deficiencies are and whether they are insured elsewhere (the risk may be adequately covered by other policies, e.g. a motor policy).

The insurance has to be effected in such a way that the employer is satisfied with the policy and the insurer chosen. The employer may have difficulty in agreeing the choice of insurer and may require professional advice here.

The policy must contain a limit of indemnity 'for at least the amount stated in the Appendix to the Form of Tender'. The limit stated in the tender documents is a minimum limit. The contractor could face claims in excess of the limit in the tender since his liability at law is unlimited. It is a mistake always to insure for this minimum limit where the risks are high. The contractor remains liable whether he is insured or not. Calculating an adequate limit of indemnity under the public liability policy is a difficult task and one for which there is no scientific method. It involves the contractor, employer and their professional advisers trying to estimate what the worst possible result could be. What is the risk of a serious fire or explosion? What is the nature of the property surrounding the contract site — is it high or low in terms of value? Can a fire once started spread quickly? What are the likely business interruption losses sustained by the owners of the property if such property were to be damaged? Is there a risk of damage arising from vibration or the removal or weakening of support? Are local rivers or streams subject to the risk of pollution? Is there a high risk of personal injury, for example nearby congested roads, or is it a green field site? Are there schools in the vicinity creating the risk of trespass by children?

A further unknown factor enters into the equation. Claims often take a long time to settle and it is not uncommon to read of cases being settled where the injury or damage took place ten years beforehand. Who can predict the level of inflation and the effect on the cost of damages in the next decade?

Thus the contractor has to give serious thought to the limit of indemnity selected under the public liability policy. One thing is perfectly clear — that the limit of indemnity bears no relation to the contract value although sometimes this is the figure used. Small contracts can produce large claims. At 1986 values a limit of £1 million for any one occurrence should be regarded as an absolute minimum.

- The policy has to indemnify the employer in addition to the contractor. The intention is that the employer has the benefit of the contractor's policy so that in the event of the employer being sued he looks for his protection under the contractor's policy. The employer is only covered provided the contractor would be covered as well if the claim were to be made against him. In other words the employer gets no better cover than the contractor. For that reason it is sensible and logical that the employer should examine the contractor's policy carefully, paying particular attention to any limitations or exclusions in that policy. For example it is not unusual for contractors to insure the public liability risk arising out of the use of motor vehicles and mobile plant under a motor or plant policy. Thus such risks would be excluded from the public liabilty policy. The employer, therefore, will have to be satisfied that the motor or plant policies suitably extends to indemnify the employer against public liability claims as required by Clause 23.

The employer may very well have his own annual public liability insurance arranged in his own name. That being the case then the employer may be insured twice in certain circumstances. If so, he will have to decide whether to regard his own public liability policy as the primary measure of protection or whether to regard the contractor's public liability policy as applying first, relying on his own public liability policy as a 'stop-gap' protection in the event of any deficiencies or failure in the contractor's policy. If the employer does not have his own public liability insurance in force it would be sensible to do so. The contractor's policy (upon which he relies for protection) may be limited in scope or be void for legal reasons e.g. non-disclosure of material facts or breach of conditions, or fail because the premium has not been paid.

The contractor must produce his policy and the premium receipt to the employer. Where he is required to insure for say £20 million and has more than one policy, e.g. a policy covering say the first £1 million of the risk and further policies 'topping up' this limit to a higher figure, say £20 million, all policies must be produced. The prudent employer will examine the policies carefully to make sure there are no deficiencies. This is a sensible precaution for two reasons. First the indemnity given by the contractor to the employer against claims for injury or damage is worthless unless the contractor has the resources to back up the indemnity. If the contractor is inadequately insured he may lack the financial resources to do this. Second, since the employer is treated as an insured person under the contractor's policy, he has the right if he so wishes to go directly to the policy for indemnity in the event of a claim. He thus needs to know to what extent he is insured under the contractor's policy.

5 The contractor's public liability policy

The liability of the contractor for causing injury to third parties or damage to third party property is covered by his public liability policy (sometimes called a third party policy).

Policies vary from insurer to insurer. Some specialise in contractor's insurances, others do not. In practice there is much variation in the cover given by policies and there is no such thing as a standard public liability policy.

A specimen policy is included in Appendix 1. It is intended to be typical of the type of policy which is met in practice. For ease of reference, the various parts of the policy have been given numbered section headings in the Appendix and these section numbers are referred to in the following text.

Section 1 – the opening clause

This is a typical opening clause to an insurance policy. There are a number of important points to note.

- The premium charged by insurers and the extent of cover they are prepared to give is geared to the business they are insuring. It is important, therefore, to see that the business is adequately described. A policy describing the insured simply as a building contractor, for example, would not cover him for civil engineering work on bridges. A policy showing the insured as a painting contractor would not cover him for demolition work.

- Insurers obtain their underwriting information and base their decision whether to accept the risk or not and the amount of premium upon information supplied in the proposal form. The proposal form forms part of the contract of insurance and

CONTRACTOR'S PUBLIC LIABILITY POLICY

any inaccuracies in it give the insurers the right to void the policy if they so wish. Hence it should be completed with great care and a copy of it always retained for future reference.

- The premium is obviously the consideration for which the insurers agree to issue the policy.

Section 2 – the insuring clause

This is the insuring clause of the policy setting out the cover. The following are the important points to note.

- The cover is subject to the other terms, exceptions, limits and conditions in the policy set out later on. In other words the document has to be read as a whole.

- The insurers agree to indemnify the insured against his legal liability. In other words they will stand in his shoes, negotiate on his behalf and pay any damages that may be necessary. The *insured* is usually a limited liability company or partnership. Many policies extend the definition to include directors and employees of the insured so that they too are protected by the policy for their activities in connection with the business.

- The cover is to pay all sums which the insured shall become *legally liable* to pay. The policy is not limited to claims in negligence, for example, but covers all ways in which the insured 'shall become legally liable'. As explained in Chapter 2, this includes all forms of liability such as negligence, nuisance, trespass, strict liability, liability under statute and contractual liability. It is a *legal* liability policy, not one covering moral liabilities.

- The promise is to pay any damages awarded against the insured, that is the amount of compensation to which the claimant is entitled. (Most claims are settled by negotiation. Few go to court.)

- The damages must relate to the claimant's injury or damage and must be awarded 'in respect of' that injury or damage. Thus the policy will protect the insured against his liability for injury or damage sustained directly by the claimant plus any consequential losses which may flow from that injury or damage. In other words, if a claimant's house is badly damaged and he incurs costs in staying in an hotel whilst the repairs are carried out, the policy will pay for the damage to the house plus the hotel expenses and any other consequential losses flowing from the damage.

- The policy protects against liability for bodily injury to third parties which includes death, illness, accident and disease. (Sometimes a wider definition may be used which includes also false arrest, detention, false imprisonment, defamation and invasion of privacy.)

- The policy protects against liability in respect of loss or damage to *property* and the consequential losses flowing therefrom. In law, property has a wide meaning and can include material property and non-material property, i.e. patents, copyrights, designs, etc. The intention of most insurers is to pay only for damage to *material* property not non-material property (although interpreting the intention from the policy is often not easy).

    The reference to *loss* of property is to cover situations where property may be lost but not damaged. For example, contractors may fail to secure the site, negligently allowing third parties to enter and steal the property of others.

    Although the policy is intended to pay only for damage to material property some policies may be extended to indemnify the insured against his legal liability for accidental obstruction, loss of amenities, trespass, nuisance or interference with any right of way, light, air or water. Such 'interference with rights' cover began to appear as an extension to many contractor's policies during 1984 and 1985.

- The policy operates only provided the injury or damage is *accidental*. Injury or damage which is inevitable, unavoidable or deliberate is not covered.

    The intention is to construe the word 'accidental' from the insured's point of view which, in the case of a limited liability company, means the directors and senior management. It is not the intention to construe the word from an employee's point of view e.g. reckless conduct by an employee may result in damage which although not accidental from his point of view is from his employers. In practice difficulties can arise in interpretation.

Injury or damage is accidental when it is the unexpected or unlooked for consequence of an untoward event which is not expected or designed. In other words it must be fortuitous e.g. an excavator damages an underground cable because the operator did not know it was there. But if the insured knew and took no precautions it would not be accidental. The inevitable consequences of damage by vibration during pile driving, concussion damage arising out of the use of explosives and dust arising from demolition sites is not accidental.

In some policies the word 'accidental' may have been omitted. Nevertheless the intention of the insurers is only to pay for fortuitous damage. Consequently a specific exclusion clause appears elsewhere in the policy excluding damage which arises out of the deliberate acts or omissions of the insured or which is inevitable or unavoidable. (Several different wordings are in use).

- The liability and the injury, loss or damage must arise in connection with the business covered by the policy. Hence the importance of describing it accurately.

- The injury, loss or damage must occur within the territorial limits covered by the policy. For risks in the United Kingdom this will usually be described as the United Kingdom. Where work is performed abroad a special extension to the policy will be required.

- The injury, loss or damage must *occur* during the period of insurance. Once the policy has lapsed there is no cover even though the events giving rise to the subsequent injury, loss or damage may have occurred during the period of insurance. Thus structures erected negligently during the period of insurance which collapse and injure a third party after the policy has lapsed will mean that the claim when made will not be covered. At the date the third party was injured the policy was not in force. Hence contractors who arrange individual policies for each contract and allow them to lapse when the contract has been finished run the risk of being uninsured for subsequent injury or damage. If they have an annual policy it would be sensible to extend that policy to cover this contingent risk.

Section 3 – the limit of indemnity
- There is a limit up to which insurers will pay claims but not beyond. This is known as the Limit of Indemnity. It is usually a limit per 'occurrence'. It is important that insurers describe exactly in what circumstances the limit of indemnity will operate because a great deal is at risk. Thus if there is an 'occurrence' e.g. an explosion, which injures a great many people, insurers do not intend to pay more than the limit of indemnity no matter how many claimants there may be. If the limit is £1 million that represents their maximum exposure. It is £1 million for the 'occurrence' not £1 million per claimant.

    Sometimes a lower limit of indemnity may apply, e.g. for damage arising out of fire or explosion. Other policies apply a lower limit for damage to property arising out of vibration or the removal or weakening of support.

- The insurers will also pay the legal costs and expenses of litigation incurred in negotiating or defending the claim. In practice these often amount to a substantial amount. These are usually paid in addition to the limit of indemnity. In other words if the limit of indemnity is £1 million and the plaintiff's claim is settled for £1 million, insurers will in addition pay the legal costs incurred in negotiating settlement of the claim.

Section 4 – the policy exceptions
Policy exceptions appear frequently in insurance policies for a number of reasons. Some risks may be insured under other policies and therefore be excluded otherwise there is a duplication in cover and unnecessary overpayment of premium. Some risks are excluded because the insurer is not prepared to cover them unless he has further underwriting information. Other risks are excluded because they are not insurable, e.g. liability arising out of ionising radiations or contamination by radioactivity from any nuclear fuel, which is a Government responsibility.

Using the same numbering as that in the specimen policy in Appendix 1, the following brief comments explain why the exclusions appear and what they mean.

1. Injury to persons under a contract of service or apprenticeship with the insured is excluded because this risk is separately insured under an Employers' Liability policy.

CONTRACTOR'S PUBLIC LIABILITY POLICY

2. Property belonging to the insured is excluded since he should have his own property insured under his own material damage policy. Thus if a contractor has loaned property to a third party and then negligently damages it while it is still in the custody of the third party, the contractor cannot claim under his own public liability policy for damage to his own property.

    Property in the insured's charge or control is excluded because insurers take the view that it too should be insured under a material damage policy. In practice a great deal of discussion takes place on the wording of this particular exception and amendments are made. However, the exception does make it clear that when a contractor is working in a third party's premises those premises and their contents are to be regarded as not being in his charge or control for the purposes of triggering off the exception.

3. Passenger lifts are excluded because they can be insured under a separate policy.

    Vehicles and plant are excluded because they can be insured under a separate motor policy. In practice the contractor has a choice of whether to insure the public liability vehicle and plant risk under a motor policy or a public liability policy. Whilst vehicles are on the road they must be insured against third party risks because of the provisions of the Road Traffic Acts. Whilst they are on a contract site they may be covered under either policy.

4. The making good by the contractor of defective work or materials is regarded by insurers as a trade risk and therefore excluded. The policy is not intended to be a guarantee as to the skill or competence of the contractor's employees. Damage to the work or structures erected is not intended to be covered for the same reason although policies vary. A measure of cover may be given by some. Thus if the structure collapses many years after completion the contractor will probably be uninsured.

5. Contractual liability, i.e. liability assumed under agreement, is often an excluded risk under liability insurance policies. The reason for this is that whilst insurers can generally calculate from their own information and experience the risk of injury or damage arising out of tort, it is less easy when

liability attaches under contractual agreements. In the United Kingdom contracting parties are free to contract in whatever terms they wish and liability is frequently transferred from one person to another. The contractor may, therefore, assume very onerous responsibilities which require special underwriting.

The exception is qualified, however, by the words in brackets '(other than as defined in Endorsement 1)' so as to protect the contractor in respect of liability assumed under building and civil engineering contracts with the employer. The point will be explained in greater detail when Endorsement 1 is considered. In some policies the exception does not appear.
6. This exception appears because it deals with risks which are not insurable in the commercial insurance market.
7. Like exception 6, exception 7 deals with risks which are not insurable in the commercial insurance market.

Section 5 – endorsements
Endorsement 1
This Endorsement overrides and amends the exception dealing with liability under agreement. Where the contractor has entered into an agreement with a principal, i.e. the employer under the ICE Conditions, then the policy will protect him in respect of the liability he has assumed but *only so far as concerns injury to third parties or damage to third party property*. The policy thus indemnifies the contractor against his liability under Clause 22(1) of the ICE Conditions. However, it is necessary to point out that the policy is not indemnifying the contractor in respect of *all* his liability under the contract. To repeat the policy only applies to accidental injury to persons and damage to property and is subject always to the other terms and exceptions of the policy. Thus the policy will not pay for the cost of making good defective work or materials or damage to structures erected by the insured since these are still excluded perils. Nor will the policy pay for liquidated damages arising from delay. These are not risks covered by the policy in the first place.

The contractor may enter into other agreements where he assumes liability e.g. hire in of plants. This liability would not be covered by the above since the plant owner is not the principal for whom the contractor has agreed to carry out work. (Some policies provide this cover.)

Endorsement 2
The policy is in the name of the insured, namely the contractor. To comply with many contract conditions which require the insurance also to indemnify the principal (employer) Endorsement 2 appears. Thus the policy grants an indemnity to the employer and complies with Clause 23(2) of the ICE Conditions which state that the contractor's policy 'shall include a provision whereby in the event of any claim in respect of which the Contractor would be entitled to receive indemnity under the policy being brought or made against the Employer the insurer will indemnity the Employer against such claims . . . .'

Section 6 – the schedule
This is self-explanatory. It is emphasised again that the description of 'the business' must be accurate. It must embrace all the activities in which the contractor is engaged. If he does work outside those business activities he is uninsured. For example in one claim a firm of tunnelling contractors were working near a factory. The factory owner urgently required some equipment in his factory to be cut out and removed. The contractors agreed to do it. They caused a fire resulting in substantial damage and were sued. They were not protected by their policy because the work they were doing did not fall within the definition of tunnelling contractors.

Other amendments that may appear to the policy
The specimen policy in Appendix 1 is meant to be a typically average policy. In practice a number of other limitations may apply. Examples include damage to property, professional negligence, hazardous work and pure financial loss.

Damage to property
The policy may exclude damage to property arising out of the removal or weakening of support or alternatively grant the cover but subject to a low limit of indemnity.

Professional negligence
The policy may exclude what the insurer considers to be a 'profes-

sional negligence' risk on the basis that this should be insured under a professional indemnity policy. Thus the policy may exclude injury or damage arising out of faulty advice, design or specification or breach of professional duty.

Hazardous work
Certain forms of 'hazardous' work may be excluded. This may be on the basis that the insurer does not wish to cover it or on the basis that the contractor does not wish to pay the additional premium to include it. The policy may, therefore, exclude work in connection with the use of explosives, tunnelling or demolition.

Pure financial loss
The policy may contain an *extension* of cover to indemnify the contractor against claims for pure financial loss *not* consequent upon injury to person or damage to property. It will be remembered that the policy only operates if there is injury to person or damage to property plus the financial losses flowing therefrom. It does not, therefore, operate if the claim is one for a purely financial loss. Claims for 'pure financial loss' present the insurers with many problems, not least of which is the one of definition. Exactly what does financial loss mean? Since it can embrace almost any claim, e.g. breach of copyright, patent, defamation or professional negligence, insurers have to be very careful how they draft their policies.

Financial loss liability is well-established in connection with professional activities where breach of professional duty may cause a financial loss to the plaintiff. It is also a well-established head of claim for breach of contract. But in 'work' situations as opposed to 'professional' situations outside contract, the law has been reluctant to recognise pure financial loss as a head of claim in tort.

Interest in financial loss cover was aroused in 1982 following the House of Lords decision in *Junior Books Ltd* v. *The Veitchi Co. Ltd*. In that case *Veitchi* were nominated flooring subcontractors for a factory being built for *Junior Books*. They laid the floor in a defective condition and whilst this did not create any danger nor did it cause any damage it had to be replaced. *Junior Books* suffered financially as a result since they have to move

CONTRACTOR'S PUBLIC LIABILITY POLICY

machinery and their business was interrupted. There was no contract between *Junior Books* and *Veitchi* so the claim was in tort. From the speeches of their Lordships in that case it did seem that a defendant's liability for financial loss on its own could arise in tort. Consequently, interest in financial loss protection increased and some contractors negotiated extensions to their policy to give them this cover.

The decision in *Junior Books* however, has been regarded as one on its own special facts and disregarded in subsequent decisions. Liability for pure economic loss in tort is unlikely to arise unless it can be shown that the plaintiff can show there was a close relationship with the defendant and reliance placed upon him (see *Muirhead* v. *Industrial Tank Specialities* [1985] 3 ALL E R 705.

When the financial loss cover is granted as an extension to a public liability policy it is subject to many limitations, e.g. it may exclude the cost of repairing or replacing defective work, failure of a product to fulfil its intended function, breach of professional duty, infringement of plans, copyrights, patents, designs and liquidated damages. Moreover not all insurers are prepared to provide this cover.

An example of such a wording follows.

'The Company shall subject to the terms and conditions of this Policy indemnify the Insured against:

Liability for accidental Financial Loss occurring during the period of insurance (not occasioned by loss of or damage to Property)
The indemnity granted by this Extension shall not apply in respect of
(a) the first 10% or £   whichever is the greater of the damages costs or claimant's expenses which shall be retained by the Insured as his own liability and uninsured
(b) any liability which is assumed by the insured by agreement (other than liability arising out of a condition or warranty of goods implied by law) unless such liability would have attached in the absence of such agreement
(c) any liability arising from a breach of professional duty
(d) passing off or infringement of patent copyright design trademark or trade name
(e) any liability for a breach of obligation owed by the insured as an employer to an Employee
(f) the cost of repairing replacing or recalling any faulty product or material'.

Sub-contractors

The insured's vicarious liability arising out of the use of sub-contractors is covered but the policy gives no direct indemnity to the sub-contractors themselves. They are responsible for arranging their own insurance. However, the position is negotiable and where it is necessary for special contracts or to meet the requirements of contract conditions, the policy can be extended by negotiation to include an indemnity to the sub-contractors themselves.

Policy conditions

All insurance policies have conditions which must be complied with. Failure to do so may prejudice the cover. Policy conditions fall into three categories, precedent and subsequent to the policy, and precedent to liability.

- There are conditions *precedent* to the policy, a breach of which renders the policy void from *inception*, e.g. a breach of the duty of utmost good faith and a failure to disclose material information.

- There are conditions *subsequent* to the policy, a breach of which renders the policy voidable from the date of the breach. An example here would be a change in the risk so that it is different from the one the underwriter thought he was covering. The policy is operative for all claims up to the date of the change but not thereafter.

- Conditions precedent to *liability* are ones which must be observed before the insurers become liable under the policy. They do not prejudice the validity of the policy but simply prejudice the insured's right to indemnity for that particular claim. An example would be a condition requiring the contractor to notify all claims promptly to the insurers. If he doesn't, the insurers can refuse to deal with the claim. Typical conditions in a public liability policy would be:
  o A requirement that immediate notification is given to the insurers of any injury or damage or any claim. The object is to give the insurers the chance to investigate immediately.
  o A condition prohibiting the insured from negotiating any claim or repudiating liability or admitting liability. This is on

CONTRACTOR'S PUBLIC LIABILITY POLICY

the basis that if the insurers are going to pay they have the right to decide these matters themselves.

o Where the claim exceeds the limit of indemnity selected by the contractor a right to pay that limit to the contractor and thereafter withdraw from the claim. In other words if the limit of indemnity is £1 million and the claim against the contractor is for £10 million the insurers can give the contractor £1 million, pay the costs incurred so far and thereafter leave negotiation of the claim with him. They have no further involvement.

o A condition that where there is more than one policy in force (as there often is), each policy shall contribute according to a formula, e.g. if there are two policies each pays half of the claim.

o A condition requiring the insured to exercise reasonable care in the conduct of his business to prevent injury, loss or damage. The policy is not intended to be a substitute for good management or competent employees. Nor is the insured expected to take short cuts on the basis that if things go wrong the insurers will pay. In practice the condition is a difficult one to apply. The duty to take care is cast upon the insured, not his employees. Moreover, as the policy is intended to pay where the contractor has been negligent, i.e. not exercised reasonable care, the insurers cannot say they will indemnify someone against the consequences of his own carelessness provided he is not careless. That would be an absurdity. In practice, therefore, before insurers can invoke the condition they have to show that the insured's i.e. director or senior management's conduct was reckless, that is he acted with actual knowledge and recognition that a danger existed but not caring whether or not it was averted.

o A condition requiring that if the risk changes the insured must give notice in writing to the insurers.

o A condition giving the insurers the right to cancel the policy by sending thirty days notice by registered letter to the insured's last known address.

6 Accident or injury to workmen – an analysis of Clause 24

The following is the wording of Clause 24.

'24. The Employer shall not be liable for or in respect of any damages or compensation payable at law in respect or in consequence of any accident or injury to any workman or other person in the employment of the Contr~ :tor or any sub-contractor save and except to the extent that such accident or injury results from or is contributed to by any act or default of the Employer his agents or servants and the Contractor shall indemnify and keep indemnified the Employer against all such damages and compensation (save and except as aforesaid) and against all claims demands proceedings costs charges and expenses whatsoever in respect thereof or in relation thereto.'

Accidents to workmen and other people in the employment of the contractor or sub-contractors were also included in Clause 22. There is thus an overlap but presumably since Clause 24 is the more specific such matters are to be dealt with under Clause 24.

The contractor has to indemnify the employer against all damages and compensation paid by him and legal costs and expenses so incurred unless 'contributed to by any act or default of the Employer his agents or servants'.

Employees of Contractor
Presumably the intention of the clause is to deal with the contractor's *employees*, that is persons under a contract of service or apprenticeship with him. Persons in 'the employment' of the contractor may include non-employees, e.g. the main contractor may employ sub-contractors. The contractor's legal liability for accident or injury to his employees is the subject of employers' liability insurance. That policy, however, relates solely to employees, i.e.

persons under a contract of service. Clause 24 also embraces accident or injury to any *workmen* or other persons in the employment of any *sub-contractor* which is a matter dealt with by the contractor's public liability policy. The result is that the contractor looking for protection against the indemnity he has given under Clause 24 will require cover under both his employers' and public liability policies. It seems a pity that Clause 24 could not have been limited to accident or injury to persons employed under a contract of service or apprenticeship with the contractor which presumably is its real purpose.

Insurance not required
There is no requirement for the contractor to insure his liability under Clause 24. This presumably is under the mistaken impression that since employers' liability insurance is compulsory by virtue of the Employers' Liability (Compulsory Insurance) Act, such a reference is not necessary. But since Clause 24 deals with accident or injury to persons in the employment of sub-contractors, it would appear that there is no contractual obligation on the contractor to insure against his liability for accident or injury to such people. This is probably not the intention but arises from the poor drafting of Clauses 22 and 24.

In practice the contractor should be required to produce evidence of his employers' liability insurance to satisfy the employer that he has complied with the law and that there are no restrictions or limitations on the cover, e.g. excluding hazardous work such as tunnelling or demolition.

Moreover, the contract conditions should require this policy to provide indemnity to the employer as does the requirement for the contractor's public liability policy.

7 The contractor's employer's liability policy

A specimen policy is included in Appendix 2. It is numbered in sections to make the following text easier to follow. As with all other insurance policies, there is no standard policy in force and policies do vary from insurer to insurer. The policy reproduced in Appendix 2 is, however, typical of the sort that are in daily use.

An employers' liability policy protects the contractor as employer against his legal liability for accidents to or diseases sustained by employees in the courts of their employment.

Employer's liability insurance in the United Kingdom is compulsory by virtue of the Employers' Liability (Compulsory Insurance) Act, 1969. There are certain exceptions such as the nationalised industries. The intention behind compulsory insurance is that if an injured employee has a claim for common law damages against his employer compulsory insurance will see that the employee obtains compensation in circumstances where the employer (without insurance) may not be in a position himself to meet a claim.

Unlimited cover
There are certain technical regulations laid down in the Compulsory Insurance Act, but these are not relevant to the purposes of this particular chapter. Although the Act stipulates that the limit of indemnity under compulsory Employers' Liability insurance shall be £2 million any one occurrence, in practice insurers provide cover for an unlimited amount.

The Act provides that Certificates of Insurance must be issued to prove that the employer is insured and copies of that Certificate must be displayed at the insured's place of business or, where he has more than one place of business, at each place of business where he employs people.

Who is an employee?

In modern times it has become increasingly difficult to recognise who is an employee so as to create an employer/employee relationship.

Grey areas exist in deciding whether a person is engaged under a contract *of* service so as to be an employee or under a contract *for* services so as to be an independent contractor.

Over the years the courts have laid down many tests for deciding whether a person is an employee. It is, therefore, relevant to see who engages the employee, pays his wages, determines his entitlement to holidays or pays his National Insurance contributions, regulates his work and has the power to dismiss him. In the main the deciding factor has been the control test. An employee is subject to the control of his employer as to the manner in which the work is done since the employer can tell the employee what to do and how to do it. However, in the case of highly skilled employees, the control test is hardly appropriate. It can not be said, for example, that a skilled surgeon is subject to the control of his employer in the way he carries out his medical work. Nor can it be said that other highly skilled people are under the control of their employer. Thus on occasions the integration test has been used. Was the work done by the employed person integral to the employer's business or merely accessory to it? If it was integral there would be an employer/employee relationship but not necessarily so if it was merely accessory to it. But the test too is not perfect. In the construction industry where there is much use of labour gangs, labour-only contractors and other self-employed individuals, border-line cases often arise. No matter what label a person gives himself, e.g. self-employed contractor, the courts are prepared to look behind the label to ascertain the true relationship.

Section 1 – opening clause

This is the similar type clause seen in connection with the public liability policy. Again it makes the proposal for insurance the basis of the contract so that any inaccuracy may invalidate the policy. Again it is important to note that it is only 'the business' described in the schedule which is covered. A firm described as a building contractor will not be covered when operating as a civil engineering contractor.

Section 2 – employees
- The cover relates only to employees under a contract of service or apprenticeship with the contractor, that is where there is an employer/employee relationship.

    As already mentioned, labour gangs and self-employed contractors often fall into a grey area and many policies are endorsed to the effect that such persons shall be deemed to be employees under a contract of service or apprenticeship for the purposes of the policy. Thus the insured's liability for accident or injury to them is covered.

- The policy relates only to employees who are employed in Great Britain, Northern Ireland, the Ise of Man and the Channel Islands, or temporarily working abroad. Doubts can arise over the word 'temporary'. In the case of long term contracts abroad it is necessary to clear the position with the insurers. Often local legislation may dictate that policies have to be arranged with insurers in the country where the contract is being performed.

- The employer is protected only against claims for bodily injury or disease. Property damage is not covered, although in practice where an employee sustains damage to his clothing and personal effects at the same time as he is injured such damage is usually paid in addition to compensation for the injuries.

- The bodily injury or disease has to be caused during the period of insurance. The word 'caused' is used deliberately with diseases in mind. Injury consequent upon an accident is usually immediately apparent. There are some diseases, however, such as asbestosis, cancer and noise-induced hearing loss which may be caused over a long period of time and may not be diagnosed until twenty or thirty years later. Provided the disease was caused during the period of insurance the policy is operative. The employer must, therefore, be able to trace his policies and the insurers for many years back (in some cases 30 years) to prove he was insured when the disease was caused. The policies will share the claim depending how much of the disease was 'caused' during the periods the policies were in force – a difficult point to establish. If the contractor cannot prove who his previous insurers were his policy requires extending to cover *claims* during its currency for diseases caused before its inception, but not all insurers are prepared to give this cover.

- The injury or disease must arise out of and in the course of employment with the contractor. Where an employee is injured in his out-of-work activities and has a claim against his employer, the employer will be protected by his public liability policy.

- The bodily injury or disease to be covered must arise in connection with the business described in the policy. Hence the importance of having the business description correctly described.

Section 3 – the indemnity
Paragraph 1
This states that the insurers will indemnify the employer against his liability at law in respect of injury or disease to his employees. The expression 'liability at law' covers all forms of liability, i.e. the policy is not restricted to negligence. The policy is, however, a strictly legal liability policy. It is not a policy which simply makes up the employee's loss of wages whilst he is away because the employer feels he has a moral obligation to do so.

The stipulation that for employees working abroad the claim must be brought in the courts of the United Kingdom does not apply to all policies and if possible should be removed since it limits the cover.

Paragraph 2
The policy protects the insured against liability he has assumed under contracts with principals (the employer) and where necessary indemnifies the principal in like manner to the insured. Thus the principal (employer) becomes in effect an insured person under the contractor's policy. It is necessary to add here that under the ICE Conditions there is no requirement for the contrac-

tor to insure or for the employer to be a named insured under the contractor's employers' liability policy.

Clause 24 of the ICE Conditions requires the contractor to indemnify the employer 'in respect or in consequence of any accident or injury to any workman or other person in the employment of the Contractor ....'

This is a liability assumed by the contractor under agreement and the wording in the preceding paragraph caters for it.

Section 4 – legal costs

In addition to paying the damages to the injured employee the insurers will also pay the legal costs and expenses incurred in handling and negotiating settlement of the claim.

The final paragraph of Section 4 deals with a technicality which arises from the Employers' Liability (Compulsory Insurance) Act which was passed to make sure that an injured employee obtained compensation. However, there were circumstances where the insurers could refuse to pay under their policy because the insured was in breach of the policy conditions e.g. by not notifying the claim to the insurers in time. In such circumstances the injured person still has his common law claim for damages against the employer but the employer may not be in a position financially to pay them. The employee, therefore, would not get his compensation.

The Compulsory Insurance Act, therefore, limited the right of the insurer to repudiate liability in certain circumstances. For example, the insured may have failed to comply with the conditions requiring prompt notification of claims to the insurers, or he may have failed to take reasonable care to prevent injury to employees. Under the Compulsory Insurance Act, the insurers cannot take these points as a means of repudiating the claim. They must pay the injured employee.

Section 5 – the schedule

This is self-explanatory.

Section 6 – policy conditions

As explained under the chapter dealing with public liability insurance, all policies have conditions which must be complied with.

A breach of those conditions may give the insured the legal right to avoid the claim.

Other points of interest

- The person indemnified by the policy is the employer, normally a limited liability company. Most policies, however, indemnify directors, managers and employees as well.

    Such persons are at risk for two reasons. First they have a duty to their employer under their contracts of service to perform their work reasonably and competently so as not to cause loss to the employer. Thus if by the negligence of one employee a fellow employee is injured resulting in the employer paying damages, the employer having paid those damages can legally recover from the negligent employee. If the insurer pays then under the principle of subrogation he can recover from the employee – but in practice there are agreements to the effect that they will not do so.

    Whilst this is the position in law, it is doubtful whether in modern times any employer would attempt to do that or allow his insurers to do so.

    Secondly each person, even though an employee, is responsible for his or her own act of negligence. Thus if one employee negligently injures a fellow employee there is nothing to stop the negligent employee being sued personally even though the employer will be vicariously liable as well.

    By extending the policy to grant an indemnity to directors, managers and employees in addition to the insured, such persons are protected.

- Directors who are employed under a contract of service with the insured are regarded as employees in the same way as any other employee. In the case of outside directors who may not be under a contract of service the policy can be altered to treat these people as though they were directors under a contract of service.

- Exclusions in the policy may apply either because the contractor has said he does not do that type of work or is not prepared to pay the premium to include that type of work. Thus the policy may exclude, for example, demolition, tunnelling or work at certain heights or certain depths in the ground.

- As explained above, the policy protects the employer against his liability for disease *caused* during the period of insurance. Often a disease may have been caused over a long period of time and claims have been made for asbestosis, cancer, noise-induced hearing loss and similar diseases where the causation factor has spanned forty years. This has presented employers and insurers with many problems.

  In simple terms an employee suing for common law damages has three years to bring his claim, starting from the date he was injured. In the case of disease it was not possible for the employee to pinpoint the exact time.

  Thus by legislation and legal decisions the employee now has three years to bring his claim, starting from the date he knew or ought reasonably to have known of his injury. Even then the courts have a discretion to set aside any defence based on the statute of limitations if they think it is fair and reasonable to do so.

Consequently employers and insurers have faced in the last decade a mass of claims for diseases where the causation factor has spanned many years.

An employer may be sued by an employee who alleges his disease has been caused over a thirty year period. The employer has to show evidence of insurance for that thirty year period. He usually has no problem in proving evidence of insurance for the recent years but often difficulty in proving evidence of insurance for the early years. If he cannot do so then a portion of the loss falls on him, i.e. that portion of the claim relative to the years when it was caused but where the employer cannot prove evidence of insurance.

If the employer cannot trace his previous insurance history, then he may be able to buy some form of retrospective insurance. This states that for *claims made* against the employer during the currency of the policy the insurers will pay even though the cause may have been many years previously.

In some cases there may be no liability before a certain time. In deafness claims for example, the courts have established that the employer could not have been aware of the danger to health before 1963 and is not liable for deafness before then.

For contractors who frequently change their insurers to get the

CONTRACTOR'S EMPLOYER'S LIABILITY POLICY

most competitive price the subject of previous insurance history and retrospective cover is important. It is also important where companies may have been acquired years previously and no investigations were carried out by the buyer to see if previous insurance was in existence.

8 Care of the works – an analysis of Clause 20

The following is the wording of Clause 20.

'20. (1) The Contractor shall take full responsibility for the care of the Works from the date of the commencement thereof until 14 days after the Engineer shall have issued a Certificate of Completion for the whole of the Works pursuant to Clause 48. Provided that if the Engineer shall issue a Certificate of Completion in respect of any Section or part of the Permanent Works before he shall issue a Certificate of Completion in respect of the whole of the Works the Contractor shall cease to be responsible for the care of that Section or part of the Permanent Works 14 days after the Engineer shall have issued the Certificate of Completion in respect of that Section or part and the responsibility for the care thereof shall thereupon pass to the Employer. Provided further that the Contractor shall take full responsibility for the care of any outstanding work which he shall have undertaken to finish during the Period of Maintenance until such outstanding work is complete.

Responsibility for Reinstatement

(2) In case any damage loss or injury from any cause whatsoever (save and except the Excepted Risks as defined in sub-clause (3) of this Clause) shall happen to the Works or any part thereof while the Contractor shall be responsible for the care thereof the Contractor shall at his own cost repair and make good the same so that at completion the Permanent Works shall be in good order and condition and in conformity in every respect with the requirements of the Contract and the Engineer's instructions. To the extent that any such damage loss or injury arises from any of the Excepted Risks the Contractor shall if required by the Engineer repair and make good the same as aforesaid at the expense of the Employer. The Contractor shall also be liable for any damage to the Works occasioned by him in the course of any

operations carried out by him for the purpose of completing any outstanding work or of complying with his obligations under Clauses 49 and 50.

Excepted Risks
(3) The 'Excepted Risks' are riot war invasion act of foreign enemies hostilities (whether war be declared or not) civil war rebellion revolution insurrection or military or usurped power ionising radiations or contamination by radioactivity from any nuclear fuel or from any nuclear waste from the combustion of nuclear fuel radioactive toxic explosive or other hazardous properties of any explosive nuclear assembly or nuclear component thereof pressure waves caused by aircraft or other aerial devices travelling at sonic or supersonic speeds or a cause due to use or occupation by the Employer his agents servants or other contractors (not being employed by the Contractor) or any part of the Permanent Works or to fault defect error or omission in the design of the Works (other than a design provided by the Contractor pursuant to his obligations under the Contract).'

- The three paragraphs of this clause apportion responsibility for loss or damage to the works between the contractor and employer.
  - 'Permanent Works' means the permanent works to be constructed, completed and maintained in accordance with the contract.
  - 'Temporary Works' means all temporary works of every kind required in or about the construction, completion and maintenance of the works.
  - 'Works' means the permanent works together with the temporary works.

The contractor has an overall obligation to complete the works and to hand over to the employer an undamaged project. Clause 8 of the ICE Conditions states that 'the Contractor shall ... construct complete and maintain the Works'.

Where responsibility for loss or damage to the works falls upon either the contractor or the employer, both have the right to sue any person whom they feel is responsible for that loss or damage, e.g. a negligent third party. If the damage is insured the insurers are entitled to that right under the principle of subrogation.

Detailed analysis

- The contractor is fully responsible 'for the care of the Works from the date of the commencement thereof until 14 days after the Engineer shall have issued a Certificate of Completion for the whole of the Works'. (The 14 days extension is to give the employer time to arrange his own insurance.) The 14 days start from the *date of issue* of the certificate.

- If the engineer issues a certificate of completion in respect of any *section* or *part* of the permanent works *before* he issues a certificate in respect of the whole of the works, 'the Contractor shall cease to be responsible for the care of that Section or part . . . 14 days after the Engineer shall have issued the Certificate of Completion in respect of that Section or part'. Thereafter responsibility for the care of that section passes to the employer who is responsible for arranging his own insurance with insurers of his choice.

- The contractor is still fully responsible for the care of any outstanding work which he has agreed to undertake during the period of maintenance.

- Any damage loss or injury from *any cause whatsoever* to the works while the contractor is responsible for them shall be put right at the contractor's own expense other than damage caused by the excepted risks. There is some doubt as to whether 'any cause whatsoever' includes negligence on the employer's part. The intention of the drafters seems to be that it does and that the contractor will not be able to sue the employer where the employer was at fault. The intention of Clause 21 – to be discussed later – is to provide insurance for the joint benefit of contractor and employer so that whoever is responsible will be protected by insurance.

- Where 'damage loss or injury arises from any of the Excepted Risks the Contractor shall . . . make good the same . . . at the expense of the Employer' if required to do so by the engineer.

- If the contractor causes damage to the Works while carrying out his duties under Clauses 49 and 50 (dealing with repairs during or after the Period of Maintenance), the contractor is liable for that damage. (He should be insured either under his contract works policy or his public liability policy.)

CARE OF THE WORKS

- The Excepted Risks relate to common exceptions in insurance policies (except that riot is normally insured under policies issued in England, Scotland and Wales the Isle of Man and the Channel Isles but not Northern Ireland). The onus of proving that an Excepted Risks operates is on the contractor.

- Use or occupation of the Permanent Works by the employer, his agents, servants or other contractors (not being employed by the contractor) resulting in injury, loss or damage to the Works is a risk carried out by the employer.

- 'Fault defect error or omission in the design of the Works (other than a design provided by the Contractor pursuant to his obligations under the Contract)' is a risk carried by the employer. He will have his own legal rights against the persons responsible for the faulty design causing injury, loss or damage to the works (who may be insured under Professional Indemnity policies against their liability). The words 'fault defect error or omission' have a wider interpretation than 'negligent' design. Thus the exception will apply to a faulty design even if it was not a negligent one.

Although Clause 20 attempts to apportion responsibilities between the contractor and employer, in practice difficulties can arise when loss or damage occurs partly by the effect of an excepted risk and partly by one that is not.

9 Insurance of works – an analysis of Clause 21

The following is the wording of Clause 21.

'21. Without limiting his obligations and responsibilities under Clause 20 the Contractor shall insure in the joint names of the Employer and the Contractor:

(a) the Permanent Works and the Temporary Works (including for the purposes of this Clause any unfixed materials or other things delivered to the Site for incorporation therein) to their full value;
(b) the Constructional Plant to its full value;

against all loss or damage from whatsoever cause arising (other than the Excepted Risks) for which he is responsible under the terms of the Contract and in such manner that the Employer and Contractor are covered for the period stipulated in Clause 20(1) and are also covered for loss or damage arising during the Period of Maintenance from such cause occurring prior to the commencement of the Period of Maintenance and for any loss or damage occasioned by the Contractor in the course of any operation carried out by him for the purpose of complying with his obligations under Clauses 49 and 50.

Provided that without limiting his obligations and responsibilities as aforesaid nothing in this Clause contained shall render the Contractor liable to insure against the necessity for the repair or reconstruction of any work constructed with materials and workmanship not in accordance with the requirements of the Contract unless the Bill of Quantities shall provide a special item for this insurance.

Such insurances shall be effected with an insurer and in terms approved by the Employer (which approval shall not be unreasonably withheld) and the Contractor shall whenever required produce to the Employer the policy or policies of insurance and the receipts for payment of the current premiums.'

As we have seen under Clause 20, the contractor is responsible for loss of or damage to the contract works (other than damage from the excepted risks) until 14 days after the engineer has issued a certificate of completion for the whole of the works. It is prudent, however, to make the contractor insure his responsibilities so that in the event of serious loss or damage both the contractor and the employer know that a sum of money will be payable quickly under an insurance policy which will enable the contractor immediately to repair or make good such loss or damage. In the absence of insurance monies, the contractor may not financially be in the position to carry out the repairs, resulting in delays causing problems both for the contractor and the employer.

Detailed analysis
- The fact that the contractor is obliged to arrange insurance does not detract from 'his obligations and responsibilities under Clause 20'.
- The insurance has to be in 'the joint names of the Employer and the Contractor'. 'Joint names' means that both the contractor and the employer are entitled to the benefit of the policy.
- The 'Permanent Works and the Temporary Works' plus 'unfixed materials or other things delivered to the Site' must be insured 'for their full value'. In calculating the full value regard must be had to the intent of the insurance. In the event of serious loss or damage what the contractor requires is a sum of money payable quickly to enable him to get on with the contract. In the event of total or near total destruction of the Works say by fire or by some other catastrophe, he must have regard to increased cost and inflation. It would be incorrect simply to insure for the contract value. Suppose the contractor is carrying out a contract worth £10 million and expected to last for three years. Suppose the Works are completely destroyed by fire shortly before handover. Allowing for the inevitable delays such as, for example, removing debris it may be some time before the contractor starts again. Certainly it is unlikely that the three year contract will be completed at the original figure of £10 million. What the insured parties require is a sum of money to enable the contractor to rebuild

the damaged works bearing in mind the increased cost. Thus in calculating the sum insured, the contractor has to estimate for the future the increased cost of rebuilding. Although some contract works policies have a small inflation factor built into them, as will be seen when looking at the contract works policy, the prudent contractor should always make sure that his sum insured represents the correct 'full value' and the prudent employer would be wise to see that he does so.

In cases where it is impossible to visualise a total loss then it is permissible for the contractor to insure not on a 'full value' basis but on a 'first loss' basis subject to Clause 21 being altered to permit this. In other words the contractor calculates what the maximum loss is likely to be and insures for that figure. This may occur when the works are to be built in sections over a wide area and it is not possible to visualise total loss of all the sections.

'Constructional Plant' must also be insured for its full value i.e. taking into account wear and tear. This definition is slightly misleading since it means 'all appliances or things of whatsoever nature required in or about the construction completion and maintenance of the Works . . .' and this includes temporary buildings such as workmen's huts. These, incidentally, are a constant source of claims because of damage by fire.

- The cover must be 'against all loss or damage from whatever cause arising (other than the Excepted Risks)'.

- The insurance must be in force 'for the period stipulated in Clause 20(1) and . . . for loss or damage arising during the Period of Maintenance from such cause occurring prior to the commencement of the Period of Maintenance'.

- It must cover 'loss or damage occasioned by the Contractor . . . in complying with his obligations under Clauses 49 and 50' (Maintenance and Defects).

- The contractor is not obliged 'to insure . . . against repair or reconstruction of any work constructed with materials and workmanship not in accordance with the requirements of the Contract unless the Bill of Quantities shall provide a special item for this insurance'. It is not clear exactly what these

INSURANCE OF WORKS

words mean but presumably they mean that insurance is not required against the cost of making good faulty or defective workmanship or materials since these are excluded risks under nearly all policies. (But consequential damage may be covered as explained in the next chapter.)

- The employer has to approve the terms of the policy and the insurer 'and the Contractor shall whenever required produce to the Employer the policy or policies of insurance and the receipts for payment of the current premiums'.

How the employer knows which insurers to approve is not easy. Presumably he will regard insurers who are Members of the Association of British Insurers (previously the British Insurance Association) as acceptable and if in doubt about others take advice.

Production of the policies to the employer is a sensible requirement. The values at risk in many contracts are substantial and it is prudent for the employer to make sure that he and the contractor are adequately protected. In many cases this responsibility is performed on behalf of the employer by the consulting engineer. It is an onerous responsibility, given the complexity of insurance policies, and one where the consulting engineer would be prudent to seek outside professional assistance unless he has his own insurance expert on his staff. If he approves a policy which subsequently turns out to be defective, causing a loss to the Employer, he runs the risk of being sued for professional negligence.

59

10 The cover given by a contractor's 'all risks' policy

Policies do vary from insurer to insurer. One in typical use is reproduced in Appendix 3. It is numbered in sections to make it easier to read the text that follows. The policy is written on the basis of 'all risks of loss or damage'. This is a misleading form of shorthand used in the insurance industry and does not mean that every loss or damage is covered by the policy.

Section 1 – damage to the works
This sets out the circumstances in which damage to the works is covered.

- The insured must have paid or have agreed to pay the premium.
- The company will indemnify the insured against loss or damage to the works by repair, reinstatement or replacement.
- To comply with the contract conditions the insured will be the contractor and the employer.
- The indemnity relates to loss or damage and all forms of loss or damage are covered. (A number of exceptions to be considered later limit the cover.) Arguments can arise as to whether the insurers require the Works to be *physically* damaged and the meaning of 'damage'. If a hole is dug and the sides collapse or it is flooded is the hole damaged?
- The property insured is described in detail in the schedule to the policy.
- The loss or damage must occur during the currency of the policy. It follows that once the policy lapses there is no cover.
- The loss or damage must occur on sites within Great Britain,

Ireland, Northern Ireland, the Isle of Man or the Channel Islands, although in practice the policy can be extended to apply to damage in other parts of the world.
- The policy is limited to loss or damage on or adjacent to the contract site and a number of extensions widen this cover as will be explained in the following section. The contractor is responsible for loss or damage to goods and materials off the site and whilst in transit and the policy can be extended to give this cover. He may also have a responsibility to insure goods and materials the ownership of which may have been vested in the Employer under Clause 54 before delivery to the site.

Section 2 – extensions

1. *Transit – loss or damage to the Works or materials while in transit is covered, other than for transport by sea or air because of the much heavier risks involved. Loss or damage to any mechanically propelled vehicle operating under its own power is not covered because the contractor will have this risk insured under his plant or motor policies.

    Employees' tools and effects are excluded because to include these would result in many small claims which would erode the premium.
2. By including the principal (employer) as an insured party under the contract the policy complies with Clause 21 requiring the contractor to insure in the joint names of the employer and the contractor.
3. Extension 3 dealing with architects', surveyors' and consulting engineers' fees is self-explanatory. The insurers will not pay the fees incurred in assisting the insured to prepare the claim against the insurers themselves since they see no logic in financing claims against themselves.
4. Many expensive claims are paid under removal of debris. It is not only the cost of removing debris from the site which is covered but also the cost of removing to another dumping ground which may be some distance away. Property remaining on the contract site may have to be shored or propped up.

---

* Note: numbers used in Sections 2, 3 and 5 correspond to those used in the specimen policy in Appendix 3.

5. Materials and goods whilst not on the site of the contract but intended for inclusion in the Works and for which the contractor is responsible are covered whilst they are stored off site. Under Clause 54 the property in certain goods and materials intended for the site vests in the employer but the contractor is responsible for any loss or damage to them and must insure them under Clause 54 against damage from 'any cause'. (The policy does not cover damage from 'any cause' since it has limitations and exclusions.)

Section 3 – exceptions

These deal with injury, loss or damage in circumstances where the insurers are not prepared to pay.

1. The insured's retained liability or excesses normally appear in policies of this type. They are intended to cut out the small claims and sometimes insurers impose high excesses when there is an adverse feature of the risk or to make the contractor take care.
2. The object of the contract works policy is to cover the risks of loss or damage during the construction period. Once the contract has been completed and the works have been handed over to the employer the risk passes to the employer and it is his responsibility to arrange his own insurance. Care is necessary to ensure that there is no shortfall in cover in this change-over period.
3. The intention of the contract works policy is to cover loss or damage during the construction period. Once that period has finished and practical or substantial completion operates the policy ceases to apply (except to pick up any liability for loss or damage under the terms of any maintenance period or defects liability period). Once the certificate of completion has been issued, the risk of loss or damage passes to the employer and it is his responsibility to arrange his own insurance. (Note that the specimen policy does not automatically give cover for the 14 days after the date of issue of the certificate – it is necessary to negotiate it.)

   The exception of 'taken into use by the owner, tenant or occupier' is important. Once there is *use* then the risk changes; the underwriter, therefore, wishes to know. Use or

CONTRACTOR'S 'ALL RISKS' POLICY

occupation is an excepted risk under Clause 20(3). The exclusion may be modified to permit use or occupation provided all the facts are disclosed to the insurer and the appropriate premium paid.

Fourteen days after a certificate of completion has been issued the insurance ceases, other than to the extent that the contractor is responsible under the terms of any maintenance or defects liability clauses.

4. There are six items contained in this exception.

(a) Damage to any *existing* structure is not covered because it does not represent the contract works. Border line cases do arise in which case there is nothing to stop an agreement being reached with the insurers to regard such existing property as the contract works for the purposes of the policy and have the sum insured increased accordingly. Thus damage to it will be covered on an 'all risks' basis regardless of fault.

Normally, however, the position is that if the contractor damages the existing property whilst carrying out the works, and this is caused by his negligence or the contractor is liable under Clause 22, the employer will have the right to sue the contractor who will obtain protection under his public liability policy.

(b) Deeds, bonds, bills of exchange, etc., are excluded because these risks are separately insured under a money policy and in any case it is doubtful whether the items constitute the permanent or temporary works.

(c) This item removes the mechanical vehicle risk from the policy because the contractor may have these risks insured under a separate motor or plant policy. If an item of plant is used solely as a tool of trade on site then the policy applies.

(d) Despite the fact that Clause 21 requires it to be insured, damage to construction plant due to its own breakdown is excluded because there is a more appropriate form of insurance to cover this risk. Such insurance is expensive and since constructional plant may be subject to rapid deterioration and damage unless subject to regular maintenance, machinery breakdown insurance is not available unless the insurers can be satisfied as to the contractor's maintenance arrangements.

(e) Vessels or craft made to float in or on water are excluded because these can be insured more adequately in the marine insurance department.

(f) This exception dealing with property for which the contractor is relieved of responsibility by conditions of contract is important. Thus even though the employer is insured as a joint named party under the policy he will have no protection in respect of loss or damage arising from the Excepted Risks. Many of these are not insurable by the commercial insurance market but two may be. Damage arising out of use or occupation by the employer, his agents or servants or other contractors is a risk which the insurers may be prepared to include in the policy when they have full details. The same may be true of loss or damage arising from fault, defect, error or omission in the design of the works, a subject which will be mentioned in greater detail later. Unless some thought is given to providing the employer with some measure of cover for these Excepted Risks at the time the insurance is arranged, he remains uninsured unless he makes his own arrangements elsewhere.

5. Insurers will not pay for the cost of repairing or replacing property which is defective in materials or workmanship nor for the cost of upkeep or normal making good since they regard these as trade risks to be carried by the insured himself. The exception is usually qualified to make it clear that the policy will pay for damage to other property which results as a consequence of the defective work or materials, e.g. a defective part of a roof falls and damages other parts of the work which are not defective. Insurers will not pay for the cost of repairing or replacing the defective part but will pay for the resulting damage.

6. Damage by wear, tear, rust and mildew is excluded because insurers again regard it as a trade risk and not insurable. The exclusion is usually qualified to make it clear that other parts of the works which are free from such defects but are damaged are covered – the same principle mentioned in point 5. The remainder of the exclusion dealing with loss or damage due to fault, defect, error or omission in design, plan or specification of the works presents difficulties in practice. The intention behind the exclusion is clear. Insurers are prepared to cover

CONTRACTOR'S 'ALL RISKS' POLICY

the risk of loss or damage to the works during construction but they are not prepared to cover what they regard as 'professional negligence' risks attaching to the drawing up of the designs or plans. The argument is that in these circumstances such people may protect themselves under the appropriate professional negligence policies. Thus any design mistake by the engineer causing damage to the works triggers off the exclusion and the policy does not operate. In some cases the exclusion is negotiable and is altered so that although the defectively designed 'bit' or 'part' which is itself defective in design is excluded other parts of the works which are damaged as a result are not. In other words part of a structure which collapses because of a bad design and damages other parts of the structures would result in the defectively designed 'part' not being covered but the other parts being covered.

The exclusion is not an easy one to apply in practice since arguments will arise about which is the 'bit' or 'part' and there have been several legal cases as to the meaning of the word 'design'. The exclusion is wider than *negligent* design i.e. it will apply if the designer has not been negligent but the design nevertheless is not satisfactory.

7. Loss or damage to property only discovered when an inventory is made is excluded because insurers only intend to pay for loss or damage which can be traced to an identifiable cause or event. They can then investigate that event to see whether it is within the terms of the policy.

    Insurers do not intend to pay for loss or damage which only occur when the contractor carries out an inventory and then discovers that property is lost or missing.

8. Delay, non-completion or consequential loss or damage of any kind is excluded because the object of the policy is to cover physical loss or damage to the works, not loss consequent upon delay. Losses incurred by the employer, e.g. loss of income or profit because of late completion following damage, may be covered under a separate policy. These are usually called advance profits policies. The contractor may be able to insure against some of the consequences of delay e.g. liquidated damages under a separate policy but it is not easy.

9-12. The remaining exceptions 9, 10, 11 and 12 deal with exclusions which are found in all insurance policies and relate to risks against which there is no commercial insurance available.

Some other exceptions

Because insurance policies are not standardised there may be additional exceptions which appear in other policies in addition to those shown in the specimen policy in Appendix 3. Other policies may exclude loss or damage arising out of cessation of work. If there is a strike or labour dispute or a problem with suppliers or delays in deliveries, the site may be left unattended or with a minimum of security. This represents an increase in risk which is why the exception appears. Where the contract involves the installation or erection of plant or machinery which has to be tested or commissioned before the employer will accept it, testing and commissioning may be excluded. This is on the basis that they represent high risks, particularly those attaching to breakdown, fire or explosion. The risks may be included, but subject to special negotiation with the insurers.

Section 4 – limits of liability

The sums insured on the works, plant and other property are set out in the Schedule (see Section 7 of the policy in Appendix 3).

Section 4 allows for the sum insured to rise to 120% of the estimated contract price plus the value of free materials.

The final paragraph allows for the sum insured to be reinstated should there be a claim. Thus if property is insured for £2 million and there is a claim for £1 million, the sum insured is not reduced. It automatically reverts to £2 million.

Section 5 – general conditions

Like all insurance policies there are conditions which have to be complied with before the insured is entitled to the benefit of the policy.

1. This is a standard condition in most policies and is self-explanatory.

Take care
2. The insured is required to *take all reasonable precautions* to prevent loss or damage since insurance is not intended to be a

substitute for good management and supervision. The theory is that the insured proceeds in exactly the same way whether or not he is insured. Insurance is not a means of taking short cuts – although in practice, of course, this happens.

Should the contractor, however, fail to take reasonable precautions to prevent loss or damage he runs the risk of breaching the terms of the policy.

It is the insured who has to exercise reasonable care, not his employees. Carelessness on the part of employees is one of the risks against which the employer seeks insurance. Who, therefore, is the insured? According to a Scottish decision, this is the board of directors. Thus in the case of a small firm of contractors where the directors themselves are aware of risk and have failed to take precautions to avoid it the insurers may be able to prove a breach of this condition. It becomes less easy in the case of a large firm where the board of directors may be well removed from the performance of the actual contract.

Change in risk
3. This condition requires any change in risk to be notified to the insurers and the insured in the meantime to take such additional precautions as are necessary. It represents an onerous condition and the insured must be careful to remember it and comply with it. It is there because the insurer calculates his premium and risk assessment on the facts known to him and if the risk changes he wants to know. It may affect his premium, his assessment of the risk and how much of the risk he decides to carry himself. Insurers often reinsure parts of their risk to keep their own exposure within reasonable boundaries.

Observing terms of policy
4. This condition is again a typical condition which makes the original proposal for insurance the basis of the contract so that if anything is wrong or inaccurate in the proposal the policy is voidable.

The condition also reinforces the fact that the insured has to comply with the terms and conditions of the policy before he is entitled to indemnity.

Premium adjustment
5. This condition is a typical clause dealing with the adjustment of premium. Most premiums are calculated by applying a rate per cent to contract values at the beginning of the year and at the end of the year the insured declares the correct values when they are known to him. An adjustment of premium is then made and the insured is either entitled to a return of premium or charged an additional premium.

Section 6 – claims conditions

These conditions should be read carefully. Note particularly that in the event of some event occurring which may give rise to a claim, the insurer has to be told in writing with full details. This is to enable him to investigate the occurrence whilst the facts are still fresh in the minds of all people concerned.

Notification to the police is necessary in the case of theft, loss or wilfull damage. In the case of loss or damage not notified to the insurers within three calendar months there is no cover under the policy.

Condition 5 allows arbitration if there is a dispute as to the *amount to be paid* under the policy. It does not arise where there is a dispute as to the words of the policy itself. Such arbitration clauses which did exist in the past were criticised on the basis that it was unfair to the insured and if he wanted the matter to go to court on the interpretation of the policy then he should have the right to do so and not be forced to go to arbitration. Insurers, therefore, restricted their arbitration clauses to matters of quantum only.

All the terms and conditions of the policy have to be understood if the insured is to comply with them. In practice the terms and conditions of the policy are frequently breached but insurers normally deal reasonably with most claims.

Section 7 – the schedule

This is largely self-explanatory, but some points require special attention.

The insured will normally be shown as the contractor (particularly if he is relying on an annual policy which has to cover more than one contract). To comply with the ICE Conditions of Contract the employer obtains indemnity under the policy under the heading of indemnity to principals discussed earlier, which for practical reasons insurers regard as tantamount to insurance in the joint names.

Where a single policy is being issued to cover the works (rather than relying on the contractor's annual policy), then the insured would be shown as the contractor and the employer.

The period of insurance must cover the full period of the contract. Where the contractor is relying on an annual policy then it should be renewed each year.

The business should be correctly described to cover all the activities, a point which has been mentioned several times previously.

The description of the contract works insured is self-explanatory and includes free issue materials. It is important to see that the sums insured are correct and are sufficient to give the contractor an amount of money to replace any work lost or damaged. To select the estimated contract price applying at completion of the contract is dangerous. Variations in the contract plus inflation during the period of the contract may make this figure inadequate. As the contract moves to a close and there is a build-up of values, what is required is a sum of money to enable the contractor to recomplete the contract should there be a total loss. The policy provides some inflation factor by stipulating in the heading of limits of liability that the sum insured under item 1 shall not exceed 120% of the estimated original contract price, including the value of free materials.

11 Remedies on contractor's failure to insure – an analysis of Clause 25

The following is the wording of Clause 25.

'25. If the Contractor shall fail upon request to produce to the Employer satisfactory evidence that there is in force the insurance referred to in Clause 21 and 23 or any other insurance which he may be required to effect under the terms of the Contract then and in any such case the Employer may effect and keep in force any such insurance and pay such premium or premiums as may be necessary for that purpose and from time to time deduct the amount so paid by the Employer as aforesaid from any monies due or which may become due to the Contractor or recover the same as a debt due from the Contractor.'

The Contractor has to arrange insurance against loss or damage to the works (Clause 21) and arrange public liability insurance (Clause 23). In both cases the insurances must be affected with an insurer approved by the employer and the terms of the policies must be approved by the employer. When required to do so by the employer the contractor must produce the policies and the receipts for payment of the premiums.

The production of the policies to the employer is a sensible requirement since the employer is entitled to benefit under both policies and clearly requires to see the extent of the cover he has and also to make sure that the policy contains no limitations or exclusions which leave the employer unprotected.

Clause 25 says that if the contractor shall fail to produce to the employer satisfactory evidence that such insurances are in force (or any other insurances which the contractor may be required to take out) the employer may take out such insurances and deduct the premiums from amounts due to the contractor. The clause refers to 'satisfactory evidence' that the insurances are in force. It is recommended that in all cases these are the policies themselves or certified copies of them. That is the only way the employer can really know the terms and conditions of the policies. The contractor or his insurers or brokers will produce certificates or letters to the effect that certain insurances are in force. These certificates or letter are never satisfactory since they usually state that the insurance which is in force is 'subject otherwise to the terms exceptions limitations and conditions of the policy'. In effect this is of little use to the employer since there may be terms, exceptions, limits and conditions which severely limit the cover and to which he would have objected had he known. It is in fact difficult for the persons issuing these certificates or letters to do anything else other than qualify them in the way described. The certificates or letters can hardly set out every word of the policies. However, if the certificates or letters confirmed that the insurances complied with the contractor's obligations under Clauses 21 and 22 – without making such a qualification – this would be untrue and could expose the firm issuing the certificates or letters to claims for damages for negligent misstatement if the employer who placed reliance upon them sustained a loss because the insurances were inadequate.

Should the contractor fail to produce 'satisfactory evidence' of insurance to the employer there are practical problems in the employer's way in attempting to arrange such insurances himself as allowed by Clause 25. Before insurers issue policies they require much underwriting information to enable them to assess the risk and the premium. This would include, for example, details of how long the contractor has been in business, his experience of the type of work involved, details of any previous insurances held by him and the details of any claims paid under those policies for the preceding five years and how much they cost. Claims experience under previous policies is one of the important factors taken into account by insurers in assessing the premium for future risks. Such knowledge is unlikely to be available to the employer who would, therefore, have great difficulty in effecting the joint names insurances required. However, it would be possible for the employer to arrange insurances in his name protecting his interest in the works and covering his legal liability risk and to charge that premium to the contractor.

12 Sub-contractors and the sub-contract agreement

In respect of claims for injury to persons or damage to property the main contractor gives an indemnity to the employer under Clause 22 of the main contract agreement. That indemnity is wide enough to include the activities of sub-contractors. Thus the main contractor makes himself responsible to indemnify the employer even for injury or damage which arises from the sub-contractor's activities.

The main contractor may also be sued by third parties for injury or damage where the claims arise from the sub-contractor's activities.

Most building and engineering contracts, therefore, contain an indemnity clause in the sub-contract agreement whereby the sub-contractor indemnifies the main contractor against certain forms of injury, loss or damage.

The following is the wording of Clause 12(1) in the Federation of Civil Engineering Contractor's Form of Sub-Contract (Revised September 1984).

> '12. (1) The Sub-Contractor shall at all times indemnify the Contractor against all liabilities to other persons (including the servants and agents of the Contractor or Sub-Contractor) for bodily injury, damage to property or other loss which may arise out of or in consequence of the execution, completion or maintenance of the Sub-Contract Works and against all costs, charges and expenses that may be occasioned to the Contractor by the claims of such persons.
>
> Provided always that the Contractor shall not be entitled to the benefit of this indemnity in respect of any liability or claim if he is entitled by the terms of the Main Contract to be indemnified in respect thereof by the Employer.
>
> Provided further that the Sub-Contractor shall not be bound to

indemnify the Contractor against any such liability or claim if the injury, damage or loss in question was caused solely by the wrongful acts or omissions of the Contractor, his servants or agents.'

- It will be noted that the indemnity applies for injury, damage or loss which arises 'out of or in consequence of the execution completion or maintenance of the Sub-Contract Works ....'

- The contractor is not entitled to the sub-contractor's indemnity if he, i.e. the main contractor, is entitled to an indemnity from the employer under the main contract.

- The sub-contractor is not bound to indemnify the contractor where injury or damage is caused *solely* by the wrongful act or omission of the contractor, his servants or agents. The word 'solely' is important. It means that the contractor must be 100% at fault. So if there is 90% of fault on the contractor and 10% on the sub-contractor, the sub-contractor still has to indemnify the main contractor for the full amount.

- The contractor under Clause 12(2) passes on to the sub-contractor the indemnity which he received from the employer under the main contract.

- The sub-contractor will need to be insured against his liabilities for both injury to his own employees and injury or damage to third parties. He will, therefore, effect Employers' Liability and Public Liability insurances. Such insurances will also indemnify him against the liability he has assumed under Clause 12.

Insurance

The following is the wording of Clause 14(1) of the sub-contract, which deals with the sub-contractor's insurances.

'14. (1) The Sub-Contractor shall effect insurance against such risks as are specified in Part 1 of the Fifth Schedule hereto and in such sums and for the benefit of such persons as are specified therein and unless the said Fifth Schedule otherwise provides, shall maintain such insurance from the time that the Sub-Contractor first enters upon the Site for the purpose of executing the Sub-Contract Works until he has finally performed his obligations under Clause 13 (Maintenance and Defects).'

It is suggested that Part 1 of the fifth schedule should require the sub-contractor to effect Employers' Liability, Public Liability and motor insurance for his own benefit and – if he is not to have the benefit of the contractor's policy – insurance of the sub-contract works, constructional plant, temporary works materials and other objects belonging to the sub-contractor.

The contract works

The main contractor has an obligation to insure the contract works under Clause 21 of the main contract. This insurance will be in the joint names of the contractor and employer. Such insurance will include the sub-contract works but only the employer's and contractor's interest in the sub-contract works. The sub-contractor is not a named party under that policy.

The following is a provision of Clause 14(2) of the sub-contract agreement.

> '14. (2) The Contractor shall maintain in force until such time as the Main Works have been completed or ceased to be at his risk under the Main Contract, the policy of insurance specified in Part II of the Fifth Schedule hereto. In the event of the Sub-Contract Works, or any Constructional Plant, Temporary Works, materials or other things belonging to the Sub-Contractor being destroyed or damaged during such periods in such circumstances that a claim is established in respect thereof under the said policy, then the Sub-Contractor shall be paid the amount of such claim, or the amount of his loss, whichever is the less, and shall apply such sum in replacing or repairing that which was destroyed or damaged. Save as aforesaid the Sub-Contract Works shall be at the risk of the Sub-Contractor until the Main Works have been completed under the Main Contract, or if the Main Works are to be completed by sections, until the last of the sections in which the Sub-Contract Works are comprised has been completed, and the Sub-Contractor shall make good all loss of or damage occurring to the Sub-Contract Works prior thereto at his own expense.'

It will be noted from the last sentence that the sub-contract works are at the risk of the sub-contractor until the main works have been completed under the main contract. Thus the sub-contractor is responsible for any loss or damage to the sub-contract works. Logically, therefore, he will arrange his own insurance in his name. This is very important where the sub-contractor may have completed his work very early, left the site yet where his works are at risk for a long time until the contract has been completed. For example, a fencing contractor erects fencing for a new motorway which will take years to complete before handover.

Clause 14(2) is confusing in the sense that it states that 'in the event of the Sub-Contract Works or any Constructional Plant, Temporary Works, materials or other things belonging to the Sub-Contractor being destroyed or damaged in such circumstances that a claim is established in respect thereof under the said policy then the Sub-Contractor shall be paid the amount of such claim.' The sub-contractor cannot be paid the amount of such claim directly by the insurers since he is not a party to the contract. Presumably it means that the main contractor having recovered the money will pass it on to the sub-contractor if he is to have the benefit of the policy – but not otherwise. That is the way insurers interpret the clause in practice.

Benefit of policy to sub-contractor

The main contractor may decide to give the sub-contractor the benefit of his policy. If that is the case then the sub-contractor should make clear that his name is added to the policy as a named insured so as to be a party to it and a beneficiary under it. He should be entitled to exactly the same benefit as is the main contractor. It follows that he should check carefully the main contractor's policies because it may have terms, conditions, limits and exclusions which are not acceptable to the sub-contractor. The sub-contractor is still responsible for loss or damage to the sub-contract works and will, therefore, be relying on the main contractor's insurances for protection. The loss will fall upon him if the main contractor's insurances are inadequate.

If the sub-contractor is not to have the benefit of the main contractor's policy, then he should effect his own insurance. In that case the fifth schedule Part II headed 'Contractors Policy of Insurance' should be left blank or marked 'not applicable'.

13 Performance bonds and some of the legal principles governing them

The contractor may get into severe financial difficulties and be unable to complete the work, or he may go into liquidation. This causes a loss to the employer. He does not get the work completed on time and runs the risk of loss of income or loss of profit. He may have to appoint another contractor to complete the work which will incur cost. Even if the contractor does not get into financial difficulties, he may in some way fail to perform his obligations under the contract, thus causing a loss to the employer.

The financial stability of the contractor and his competence to do the work are factors taken into account by the employer when deciding to appoint a contractor. Nevertheless, the employer often requires a guarantee from a third party that if the contractor should get into financial difficulties or fail to honour the terms of his contract, that third party will pay to the employer a sum of money representing the employer's losses. The third party is called the surety and binds himself with the contractor to pay losses sustained by the employer, up to an agreed amount, if the contractor does not perform and observe all the terms and conditions of the contract.

- A bond, therefore, involves three parties:
  - the employer who benefits by the bond; if the contractor does not perform his contract correctly the bond will pay the employer's loss;
  - the contractor who has obligations under the building or engineering contract; if he does not perform those obligations he may be in breach of contract; the primary responsibility to complete the obligations is upon the contractor but if he fails

to do so (he may have gone into liquidation) then the surety will be called upon under the bond; and
o the surety is the person who jointly with the contractor guarantees to the employer that the contractor's obligations under the contract will be performed.

Bonds are not insurance

Although bonds are issued by insurance companies, bonds are not insurance. Insurance normally involves a relationship between two parties, the insured and the insurer, whereas bonds require a three-party relationship. Moreover, the contractor usually obtains no protection under the bond since the surety, if called to pay any claim under the bond because the contractor has failed to honour his obligations, has common law rights of recovery against the contractor. These are often strengthened by the provision of a written form of counter-indemnity from the contractor to the surety.

In simple terms, therefore, the bond does not relieve the contractor from any of his obligations. His obligations will be set out in the contract and he must perform them. If he does not, then the employer may sue for damages. The surety only becomes liable when the contractor has failed to meet those obligations. He then has to indemnify the employer against his proven loss up to the amount of the bond. A bond is thus an undertaking given by a surety to make payment to the employer, or otherwise guarantee performance, upon the contractor's default. A bond is a contract of guarantee and therefore has to be evidenced in writing and must be under seal. This is because the employer gives no legal consideration since it is the contractor who pays the premium for the bond (although the cost of the bond will be reflected in the contractor's price).

Often, years ago, the surety entering into the bond was a friend of the contractor and usually did not understand the risks he was taking. The courts, therefore, have always adopted a sympathetic approach to claims against a surety and have released him from his liability if the contractor's obligations have materially changed, for example, if an extension of time was granted or more extensive work was required. That is why, in order to protect the employer, the ICE Bond has a requirement that any change in the contract or extensions of time will not invalidate the bond. This will be considered in more detail later.

Once the bond has been sealed by the contractor and his surety, it is delivered to the employer and remains in force until the contractor has completed his obligations. The surety is, therefore, at risk for as long as the employer holds the bond. This could be for a substantial period of time since defects in the works may not manifest themselves until many years after the contract has been completed. However, many bonds stipulate the date upon which they will cease, e.g. upon the issue of the maintenance certificate at the end of the maintenance period.

Bonds may be issued by insurance companies, banks or private sureties. Private sureties, however, are unsatisfactory from the employer's point of view as they can become bankrupt or die.

Form of bond

The form of bond used in the ICE Conditions is shown in Appendix 4. The contractor and surety are 'held and firmly bound' to the employer. This is a joint and several liability and the employer may sue one or both of the parties for the amount of his loss.

The bond recognises that the contractor has agreed with the employer to enter into a contract for the construction, completion and maintenance of the works and that if the contractor duly performs and observes the terms of the contract the bond is null and void. If the contractor does not perform and the surety satisfies the employer's loss, the bond is further null and void. But if the contractor does not fulfil his obligations, the bond will remain in force until he does so.

The bond also provides that 'no alteration in the terms of the said Contract made by agreement between the Employer and the Contractor or in the extent or nature of the Works to be constructed completed and maintained thereunder and no allowance of time by the Employer or the Engineer under the said Contract nor any forbearance or forgiveness in or in respect of any matter or thing concerning the said Contract ... shall release the sureties/surety from liability under the bond.' Without this requirement the common law position would release the surety as was explained previously, i.e. changes in the risk do not bind the surety.

The employer has to prove his loss before the surety pays. The bond is not an 'on first of our demand without any conditions of proof' bond. These on demand bonds, which are common in the Middle East, are really a form of blank cheque in favour of the employer which can be cashed in at any time the employer wishes without the contractor being in default of his obligations under the contract.

A surety may be released from his obligations if there is a dishonest failure on the employer to disclose material information which would affect the surety's decision whether to act or not. Since bonds are not contracts of insurance, there is no general duty on the employer to disclose all material information as would be the case under an insurance policy. A bond may, however, be invalid if it is obtained by fraud or dishonesty by the employer (but not by the contractor). A surety is expected to estimate the risk involved and reflect this in the premium he charges, but the employer must not conceal any fact material to the risk which the surety would not reasonably have expected to exist, for example, that the employer knows that the tender price is far too low in comparison with other tenders because of some mistake by the contractor.

Counter-indemnities

The surety having paid has a common law right to recover his losses from the contractor. The surety may strengthen his common law position by requiring the contractor to give a specific written indemnity. Since this may be of little value if the contractor himself has gone out of business, the counter-indemnity may be obtained from the contractor's parent company or from directors of the company.

Sub-contractors

The main contractor is responsible for completing his contract and the bond will apply even if the fault is that of the sub-contractor. If there is a risk that one of the domestic sub-contractors may fail, the main contractor may wish to consider calling for bonds from the sub-contractor.

Who issues the bond

Bonds are issued by banks and insurance companies. There are advantages and disadvantages attaching to both methods. Banks regard bonds as an extension of their line of credit to the contractor. They usually know him well, much better than an insurance company would, for example. Banks can therefore issue bonds speedily but the bond when issued will affect the line of credit, perhaps prejudicing the advance of money in the future to purchase plant and materials. Banks usually pay more promptly as they are less willing to argue the merits of any claim than an insurance company, and do not make the same investigations as an insurance company would.

Before an insurance company will issue a bond it makes careful enquiries into the contractor's financial stability and his ability to perform the contract. These enquiries made by insurance companies are helpful to the employer since he knows that if an insurance company has issued a bond it is satisfied with the risk. Conversely if the insurance company refuses to issue the bond then the employer should ask himself whether he has chosen the right contractor.

The bond is usually for 10% of the contract price. Higher amounts are requested in practice. The employer has to take a view on whether he feels that 10% is sufficient.

Underwriting requirements

The following are typical of the questions an insurance company asks to ascertain the financial stability of the contractor and his competence to perform the contract:

- What is the contractor's financial standing? Audited balance sheets, trading and profit and loss accounts are required covering at least the last three years. If the contractor is a member of a group, the group accounts will also be required.

- Does the contractor have the financial resources to carry out the contract in addition to his existing commitments? Details are required as to the value of all outstanding contracts.

- Is the contractor competent to do the work and has he the technical ability? Has he completed similar contracts successfully in the past?

- Does the contractor have sufficient and suitable plant for the performance of this particular contract?

- What is the state of his present bank account and does it provide for any overdraft facilities? How is the overdraft secured?
- After making allowances for all contingencies, is the contract price sufficient to produce a reasonable profit in the end?
- Has the contractor or any of the directors ever required a surety to make payment under a bond in the past, or been a bankrupt or been a director of a company to which a receiver or liquidator has been appointed?
- Are the conditions of the building or engineering contract and the form of the bond reasonable in the obligations they seek to impose on the contractor?
- Is any part of the work to be sub-contracted, and if so, how is the contractor protected against the failure of his sub-contractors?

Other forms of bond

As well as the bond described, there are many other forms of bond.

- A bid or tender bond is intended to satisfy the employer that a bid is a responsible one and that if the bid is accepted the tenderer will proceed to execute a contract and any bonding requirements that the contract contains. If the tenderer fails to do so, the employer incurs loss and expense in going through a tendering procedure again and those losses would form a claim under the bid or tender bond.
- Advanced payment, progress or repayment bonds protect the employer against any monies that he has advanced being lost through the default of the contractor.
- Maintenance bonds protect the employer and are an alternative to retention monies during the maintenance period. They guarantee that once the contract works have been completed the contractor will carry out his obligations under the maintenance period or defect liability period of the contract.
- Retention bonds are bonds issued in lieu of retention monies, which may be required throughout the period of the contract or against the early release of the retention element of monthly progress payments.

14 The basis of legal liability for breach of professional duty

In the last two decades the number of claims against professional persons for breach of professional duty had increased enormously. Since 1963 the liability of professional persons has been widened by the courts. This increased liability is one of the reasons why professional negligence insurance is now so expensive.

Liability in contract
Traditionally, the liability of the professional person has been to his client under the terms of the contract for his services. One of the implied terms of the contract will be to the effect that the professional person will exercise reasonable care and skill. Sometimes the contract may contain a specific proviso as, for example:

> 'The Consulting Engineer shall exercise all reasonable skill care and dilligence in the discharge of the duties to be performed under this Contract ...'

Where there is no specific requirement to this effect the courts will imply it into the contract.

Standard of care
The standard of care expected from professional persons was laid down over 100 years ago in *Lanphier* v. *Phipos* (1838) where the judge said:

> 'Every person who enters into a learned profession undertakes to bring to the exercise of it a reasonable degree of care and skill. He does not undertake if he is an attorney that at all events you shall gain your case. Nor does a surgeon undertake that he will perform a cure, nor does he undertake to use the highest possible degree of skill. There may be persons who have higher education and greater advantage

than he has but he undertakes to bring a fair reasonable and competent degree of skill'.

In *Greaves & Co.* v. *Baynham Meikle* (1975) involving a claim against a consulting engineer the judge said:

> 'apply this to the employment of a professional man. The law does not usually imply a warranty that he will achieve the desired result but only a term that he will use reasonable care and skill. The surgeon does not warrant that he will cure the patient. Nor does the solicitor warrant that he will win the case.'

Warranties of guarantees

The liability of the professional person will be governed by the terms of the contract and it is a question of reading the contract as a whole. The contract may require the professional person to warrant or guarantee his professional skill or work. In that case, liability may arise under the terms of the warranty or guarantee, regardless of negligence. It represents an increased risk.

To decide whether a person has exercised reasonable care and skill, the standards expected of the profession as a whole at the time are used as a yardstick. Would other professionals have acted in the same way? If the answer is yes, then he has not fallen below the standard required.

Liability to third parties

Until 1963 the liability of the professional person was to his client, since it was to him that the duty of care was owed and it was the client who paid for the services. In 1963 the House of Lords in *Hedley Byrne* v. *Heller & Partners* extended liability to third parties who were not in a contractual relationship with the professional person. The House of Lords held that:

> 'If in the ordinary course of business or professional affairs a person seeks information or advice from another who is not under contractual or fiduciary obligation to give the information or advice, in circumstances in which a reasonable man so asked would know that he was being trusted or that his skill or judgement was being relied upon and the person asked chooses to give the information or advice without so clearly qualifying his answer as to show that he does not accept responsibility, then the person replying accepts a legal duty to exercise such care as the circumstances require in making his reply

and for a failure to exercise that care an action for negligence will lie if damage results.'

Owing to that decision the professional person now owes a duty of care in tort to third parties who are placing reliance upon his skill, advice or judgement. Thus in preparing plans or reports, or issuing certificates in circumstances where a third party other than the client may place reliance upon them, a duty of care will arise.

In cases involving personal injuries to third parties or damage to their property (as distinct from claims for economic or financial loss without injury or damage) the professional person has always owed a duty of care generally to others, not to behave negligently. Thus negligence, e.g. bad design of a structure causing it to collapse on passers-by, will render the professional person liable to the injured persons.

At one time the liability of the professional person to his client was solely contractual. Now the liability is both contractual and tortious. The matter is important because different limitation periods run in contract and tort. Generally a person has six years in which to bring a claim for breach of contract starting from the *date* of the *breach*. In claims in tort it is generally six years from the date of the *damage* (or in cases of personal injury, three years from the date of the injury). Thus the limitation period runs for a longer time in tort cases.

In 1985 the Law Reform Committee recommended that in cases of latent defects, i.e. damage that is hidden and cannot be discovered, such as faulty foundations, the six-year period should be extended by allowing the claimant a further three years in which to bring his claim from the date he *discovered*, or ought reasonably to have discovered, the defect. This extension is subject to a long stop which would bar claims brought more than 15 years from the plaintiff's breach of duty. Legislation to implement the recommendations is expected (the Latent Damage Bill was in draft form at the time this book was written.)

Limit of liability

It may be possible for a professional person to exclude or limit his liability to the client under the terms of the contract. The clauses would, however, be subject to the Unfair Contract Terms Act

(1977) and would have to be fair and reasonable in the opinion of the courts. It is not possible to contract out of liability for personal injuries.

While complete exclusions of liability are unlikely to be considered fair or reasonable, limitations on liability may be. There is no reason why a professional person should not attempt to limit his liability to a stipulated sum, e.g. £100,000 or a multiple of his fee. It is not unusual to find some professional firms limiting their liability to the extent of cover they have under their professional indemnity policies. To be effective, such limitations of liability must be brought to the attention of the client before the contract is agreed, so that he is aware of them and understands them. Such limitation or exclusion clauses are ineffective against third parties since such third parties are not parties to the contract.

Contractor's design and construction contracts

The preceding comments relate to professional firms engaged under a contract to provide professional services. The liability of the contractor designing and constructing is different. If a contractor agrees to design and construct a building, a client is entitled to a structure which is free from defect. Thus in the event of design faults appearing in the structure, it is no defence for the contractor to show that he was not negligent. He must comply with his contractual obligations. There is thus an essential difference between design functions undertaken by a contractor and those taken by an independent design consultant.

15 The cover given by a professional indemnity policy

As with all insurance policies, there is some variation in cover. However, a typical policy issued to consultants will agree:

'To indemnify the Insured against any claim for damages for breach of professional duty which may be made against him during the period of insurance due to any negligent act error or omission whenever or wherever the same was or was alleged to have been committed by the Insured or their predecessors in business or any employee of the Insured or their predecessors in business in the conduct of the business.'

The following points are important.

- The claim must be for a breach of *professional* duty. Claims by visiting clients who have been injured by a fall in the office, e.g. worn carpet, or by defects in refreshments supplied to him do not arise from breach of a *professional* duty. These claims are covered by a public liability policy.
- The policy is a *claims made* policy, in other words it responds only to claims made during the currency of the policy. The negligent act, error or omission giving rise to the claim may have been committed some years ago but the policy nevertheless applies (assuming the negligent act, error or omission was unknown to the insured at the time he proposed for insurance – if it was known but not disclosed to the insurers, the policy would not provide cover because of the non-disclosure of material fact).

  Once the policy lapses there is no cover, notwithstanding the fact that a negligent act, error or omission was committed during the currency. The sole test is: was the policy in force at the date of the *claim*.

- Cover relates to any negligent act error or omission. The words are important since if liability arises solely under a warranty, fitness for purpose or guarantee clause in the contract for services, liability may arise under that clause regardless of any negligent act, error or omission and there would be no cover. It may be possible to arrange wider cover, e.g. to cover 'all civil liability' or 'breach of duty' but it is not easy. Since 1985 the professional indemnity insurance market has contracted and become more difficult owing to large losses.

- There is a limit of indemnity up to which insurers will pay claims, but not beyond. This limit is usually an *aggregate* for all claims in any one policy year.

  An aggregate limit was normal practice for professional indemnity policies for many years, but some insurers are prepared to offer an any one 'claim' limit or an any one 'occurrence' limit. However, during 1985 and 1986 the professional indemnity market contracted considerably owing to increasing claims and underwriting losses made by insurers. These 'claims' or 'occurrence' limits are not now as easily obtainable.

- Insurers will pay the costs and expenses incurred with their consent in the defence or settlement of any claim. However, if when the claim is ultimately settled it is higher than the limit of indemnity, the legal costs and expenses are shared in proportion to the claim and the limit of indemnity. Thus if the limit of indemnity is £1 million and the claim is settled for £2 million, the insurers will pay only one-half of the legal costs and expenses.

- It is important to insure for an adequate limit. Claims often take a long time to settle and upwards of five years is not unusual. Regard must be had to inflation and monetary values applying at the date of settlement which may increase substantially the cost of a claim.

- The 'business' will be carefully defined in the policy, and the policy will operate only in respect of those business activities defined. Care is necessary, therefore, to see that the description is sufficiently wide and covers all the professional duties performed.

CIVIL ENGINEERING INSURANCE AND BONDING

- Work in connection with consortia and joint ventures is a matter which requires watching carefully. The insurers should be kept advised and agreement reached as to exactly what cover is provided. The contractual agreements for such work require careful scrutiny. One party may find himself carrying the responsibilities of the other party (without any fault on his part) which can have serious consequences if the other party is uninsured or goes out of business.

- If the policy is an annual policy, it may or may not exclude work outside the United Kingdom. This is because liability in overseas countries may differ substantially from liability in the United Kingdom. In addition it will be important to establish which system of law will govern any dispute, i.e. English law or foreign law. Even if work abroad is covered it may be subject to a legal jurisdiction clause in the policy, e.g. claims only covered if made in UK courts. Such jurisdiction clauses must be read with care and their meaning understood.

Extensions to the policy
It may be possible to extend the policy to give the following additional cover:

- Claims for libel or slander.

- Claims from third parties who have sustained losses arising out of the fraud or dishonesty of employees. Deliberate acts of fraud or dishonesty by employees are not intended to be covered by the policy, which refers to any 'negligent act, error or omission' i.e. some inadvertent mistake. Thus an employee may deliberately give misleading information, destroy plans, records or documents, give false certificates or permit other acts of dishonesty for which his employer may be liable.

    This extension does not cover loss of the employer's *own* money or property which is the subject of fidelity guarantee insurance. The policy is intended to cover claims by third parties, not by the insured himself.

- Claims for alleged infringement of copyright of third parties.

- Protection to employees in addition to the employer. The employer's name will usually be shown as the insured party in the policy. Legally, having settled a claim against the employer arising out of an employee's negligence, the insurers may have rights of subrogation against negligent employees. An indemnity to employees clause (or a waiver of subrogation rights against employees) gives protection to the employee.

Before insurers issue a professional indemnity policy, they insist upon completion of a proposal form asking a number of specific questions. It is necessary to repeat again that the answers to the questions must be complete and the insured must not fail to disclose any material facts that may have a bearing on the underwriter's acceptance of the risk.

It is usual for insurers to ask for completion of a declaration form (or in some cases a new proposal form) at each renewal date of the policy. The duty to disclose all material facts arises again at renewal date. Great care has to be taken in giving correct and full, information to the insurers.

It is sensible for the person signing the proposal form to consult with his directors, partners, departmental managers and senior staff before doing so to make sure that all material facts are disclosed and the proposal form has been completed correctly. It is sensible to have an item on the agenda for partners meetings or board meetings covering professional indemnity matters generally and insurance, so that anything which is relevant to professional liability or may give rise to a claim or has a bearing on the professional indemnity policy is considered immediately and if necessary discussed with the insurers. Failure to do so may result in inadvertent non-disclosure of material facts or breach of the policy conditions rendering the cover invalid.

Contractor's design and construction contracts

A professional indemnity policy is often issued to contractors to cover the liability arising from mistakes in their design department. As was explained in Chapter 14, the liability of the contractor for design mistakes extends beyond negligence. Subject to whatever the contract may say, the contractor is normally liable for his design faults whether there is negligence on his part or not. It follows that a professional indemnity policy which protects the contractor against 'negligent acts errors or omissions' in his design department does not give full protection, but insurers are not keen to grant cover going beyond negligent acts, errors or omissions. They are not prepared to 'guarantee' that the design will do what it is intended to, or cover liability under contract where there has been no negligent act, error or omission.

The contractor may discover a mistake before the contract is completed and have to put it right without there being any form of 'claim' against him by the employer. A form of 'first party' cover is sometimes given to reimburse the contractor for the costs involved. (See specimen policy in Appendix 6 under 'Special Provisions'.)

Policy conditions

There are a number of important policy conditions which must be complied with. Failure to do so may prejudice the policy. Unfortunately they are breached in practice and insurers then refuse to indemnify the insured. It cannot be too strongly emphasised how important it is to comply with them.

- The insured must not admit liability or in any other way attempt to negotiate or compromise a claim against him. All negotiations must be carried out by the insurers unless they give permission for the insured to do so.

    To avoid adverse publicity and differences of opinion between the insured and the insurers, in the case of difficult claims the insurers will not require the insured to contest any legal proceedings unless a Queen's Counsel mutually agreed upon by the insured and the insurers advises that such proceedings should be contested.

- The insured must give immediate notification to the insurers of any claim made against him or receipt of any information from any person of an intention to make a claim against him.

- If the insured becomes aware of any 'occurrence' which may subsequently give rise to a claim (even though no claim has been made at the time) he must notify that 'occurrence' to the insurers. For example, he may discover a serious mistake in the office which has not yet given rise to a claim or may have reason to believe a design is faulty.

    Most policies provide that once that occurrence has been

PROFESSIONAL INDEMNITY POLICY

reported to the insurers, any *subsequent* claim which *may* be made against the insured arising out of that occurrence (even if many years later) will be *deemed to have been made* during the currency of the policy in force when the occurrence was notified.

The interface between various policies

Some of the professional negligence risk may be covered by the contractor's public liability policy discussed in Chapter 5. Exactly how much and to what extent depends upon how adventurous that insurer is prepared to be in covering the 'professional' activities. The insurance market divides itself into specialist areas so that some insurers specialise in covering 'professional' risks and others specialise in covering contractor's public liability risks. The dividing line, however, is by no means clear and there can be grey areas. As a generalisation, many contractors' public liability policies will cover the contractor's liability for injury to third parties or physical damage to third party material property (other than the Works or structures erected) arising out of professional activities. It depends upon whether the policy specifically excludes claims arising out of breach of professional duty or arising in connection with advice, design, plans or specifications.

The position with regard to damage to the contract works is that the contractor's public liability policy does not cover this form of damage. During the construction period, the contract works are separately insured under a separate contract works policy. Damage or defects occurring after practical or substantial completion are not usually covered by a public liability policy, since insurers are not willing to guarantee, by paying such claims, that the contractor will perform the contract adequately. Normally contractors are uninsured for defects or damage arising after practical completion, except for any special 'professional indemnity' cover they may decide to buy for their design activities or any extension to the contract works policy to cover defects or damage during the maintenance period.

A summary of the position follows.

Public liability

- This policy protects the contractor against his legal liability for injury to third parties and physical damage to material property of third parties, plus any consequential losses which may flow from that injury or damage.

It will cover the 'professional' activities of the contractor for such damage, provided there is no specific policy exclusion of claims due to breach of professional duty or arising from advice, design, plan or specification.

- It does not cover damage to the contract works arising out of faulty construction either during construction or after practical completion (unless a special extension has been agreed).

- Claims for purely economic losses (unrelated to physical damage to property) are not covered by the policy. If the works do not fulfil their purpose after practical completion, causing a loss of income or profit to the employer, the contractor is uninsured in respect of these claims.

Professional indemnity
- In the case of policies issued to contractors to cover their design departments the risks described in the preceding section on public liability (first paragraph) may or may not be included in the professional indemnity policy. There will be no need to include them if they are covered by the public liability policy.

  In the case of policies issued to consultants, liability for injury to third parties and damage to third party property will normally be covered.

- In the case of contractors, the cover would include liability for damage to or faults in the contract works after practical completion – this is its main purpose. The cost of rectifying faults before completion is not covered since this is a cost down to the contractor and he cannot sue himself. The policy is intended to protect against claims made by *third parties*. However, some measure of 'first party' cover can often be included in contractors' design and construction policies to cover the costs of rectifying faults discovered during the construction period and prior to handover, where these faults are due to some neglect, error or omission on the part of the contractor's design department. (See specimen policy under 'Special Provisions', Appendix 6.)

- Policies issued to consultants have no such limitation and, to the extent that they have a liability for defects in or damage to the works before or after completion, the policy protects them.

Contract works policy
This covers damage to the contract works during construction. It will normally have an exclusion of all damage to the works arising out of design, plan or specification. The exclusion may be modified so that it is only the defective 'part' which is excluded. Thus if the defective 'part' causes damage to non-defective parts, damage to those non-defective parts is covered. Problems arise in practice in defining which is the 'part'.

16 Risk management

A sensible approach to the risks that may affect an engineering contract can eliminate some of them and mitigate the effects of others. It is logical to identify and control risk at the outset, rather than to argue about who is responsible for injury, loss or damage after it has taken place.

- Risk management entails a sensible overview of risks and falls into four stages:
  o the identification of those risks which can arise
  o an analysis and measurement of those risks to see what is involved
  o treatment of those risks so as to eliminate them or reduce their impact
  o controlling those risks or transferring them to another party, either by means of clauses in the contract conditions or by insurance.

The insurance industry has much to offer in the field of risk management. Its vast experience in dealing with the many and varied claims under insurance policies makes it uniquely qualified to give advice on how to prevent similar events in the future. It therefore has an important role to play in the overall risk management assessment to a contract at both design and construction stages. Good risk management techniques not only prevent or eliminate injury, loss or damage but also often make the project more easily insurable once the contract has been handed over and the works become operational.

Some examples of the insurance industry's approach to risk management follow.

RISK MANAGEMENT

Damage to property and loss of profit or income, or delay
The following actions may be recommended:

- investigate materials and construction methods, storage facilities, check and test for fire and explosion protection;

- examine topography and seismic information to estimate the risks of avalanche, landslip, subsidence, earthquake and tidal wave and investigate the earthquake resistance of structures;

- look at the climate and its effect on property, checking windstorm resistance and the standards adopted;

- consider the possibilities of snow and frost damage, flooding from the sea, rivers, canals or lakes or from imperfect drainage;

- consider the risks of impact by vehicles and contractors plant;

- check proximity to airfields, flight paths and to helicopter landing pads in connection with the risk that aircraft or parts of them may fall on the contract site;

- look at overall security and the risks of theft of property and vandalism;

- consider with financial executives, production and plant engineers, the effects of business interruption and financial loss or delay resulting from the risks above;

- identify key components and examine their susceptibility to loss or damage while at the maker's premises or outside warehouse, in transit to the contract site or in subsequent storage. Particular attention should be given to parts supplied from abroad. Loss or damage could mean long delays.

Public liability
Attention could be given to:

- the impact of a serious fire or explosion causing injury to people or damage to surrounding property;

- pollution or contamination of nearby land, rivers or crops;

- noise levels;

- the liability arising out of the use or employment of contractors or sub-contractors and how their work is co-ordinated;
- the presence of children – does the site have any allurements?
- is it adequately secured to keep children or trespassers out?
- contractual documents or agreements which may have a bearing on liability or its apportionment.

Health and safety
Action should be taken as follows:

- produce an effective health and safety policy, detailing the organisational hazards, standards required to meet legislation and recognised good practice and the means to monitor and control the work;
- advise on the measures to be taken during the design of the works to avoid
  o fire or explosion;
  o the harmful emission of toxics;
  o health hazards;
  o physical hazards associated with access to work places, lifting operations, pressure plant and process machinery, entry to confined space, collapse of the structures, and so on.
- diagnose any medical problems caused by technology, materials, processes or methods of work and try to reduce diseases by early advice and proper screening measures where appropriate.

Appendix 1. Contractor's public liability policy

Comments on the following specimen contractor's public liability policy are made in Chapter 5. The numbered section headings have been added to facilitate cross-referencing to Chapter 5, and would not appear in the policy.

Section 1
Whereas the insured carrying on the Business described in the Schedule and no other for the purposes of this insurance by a proposal and declaration which shall be the basis of this contract and be deemed to be incorporated herein has applied to the Company for the insurance hereinafter contained and has paid or agreed to pay the Premium as consideration for or on account of such insurance.

Section 2
Now this policy witnesseth that the Company will subject to the terms exceptions limits and conditions contained herein or endorsed hereon indemnify the Insured against all sums which the Insured shall become legally liable to pay as damages in respect of:
1. Accidental bodily injury to any person;
2. Accidental loss of or damage to property;

happening in connection with the Business and occurring within the Territorial Limits during the Period of Insurance.

Section 3
Provided that the liability of the Company for all damages payable to any claimant or any number of claimants in respect of or arising out of any one occurrence or in respect of or arising out of all occurrences of a series consequent on or attributable to one source or original cause shall not exceed the Limit of Indemnity specified in the Schedule.

In respect of a claim for damages to which the indemnity expressed in this policy applies the Company will also indemnify the insured against:

(*a*) all costs and expenses of litigation recovered by any claimant from the Insured;

(b) all costs and expenses of litigation incurred with the written consent of the Company;
(c) the solicitor's fee for representation at any coroner's inquest or fatal enquiry or in any court of summary jurisdiction.

In the event of the death of the Insured the Company will in respect of the liability incurred by the insured indemnify the Insured's personal representatives in the terms of and subject to the limitations of this Policy provided that such personal representatives shall as though they were the Insured observe fulfil and be subject to the terms exceptions and conditions of the Policy so far as they can apply.

Section 4
Exceptions
This Policy does not cover:

1. Liability in respect of injury to any person under a contract of service or apprenticeship with the insured where the injury arises out of and in the course of such person's employment or service with the Insured.

2. Liability in respect of loss of or damage to property:
    (a) Belonging to the Insured.
    (b) In the charge or under the control of the Insured but this Exception shall not apply to:
    (b1) Property belonging to any servant of the Insured; or
    (b2) Premises not owned or rented by the Insured but temporarily occupied by the Insured for the purposes of alteration or repair thereof or therein.
    (c) Caused by or through or in connection with the bursting of any economiser used in conjunction with a steam boiler vessel or other apparatus which is intended to operate under internal pressure due to steam and belonging to or in the charge of or under the control of the Insured.

3. Liability in respect of injury loss or damage caused by or through or in connection with:
    (a) Any passenger lift passenger elevator or passenger escalator owned by or in the possession of the Insured. This exception shall not apply in respect of the occasional carriage of passengers on any goods lift goods elevator or goods escalator.
    (b) The ownership or possession or use by or on behalf of the Insured of:
    (b1) Any vehicle (or machine) which is capable of self-propulsion or attached to a self-propelled vehicle and used in circumstances to which the Road Traffic Acts apply; or

APPENDIX 1

(*b2*) Any vehicle (or machine) which is insured for the benefit of the Insured under any form of Motor Insurance Policy; or

(*b3*) Any vessel or craft not specified in the Schedule under the heading of Plant.

4. The making good, replacement or re-instatement of defective work carried out or materials or goods or structures supplied erected altered or repaired by or on behalf of the Insured.

5. Liability assumed by the Insured by agreement (other than as defined in Endorsement 1) unless such liability would have attached to the Insured notwithstanding such agreement.

6. Any legal liability of whatsoever nature directly or indirectly caused by or contributed to by or arising from ionising radiations or contamination by radioactivity from any nuclear fuel or from any nuclear waste from the combustion of nuclear fuel.

7. Liability for any consequence of war invasion act of foreign enemy hostilities (whether war be declared or not) civil war rebellion revolution insurrection or military or usurped power.

In these Exceptions the expression 'vessel or craft' shall include any vessel craft or thing made or intended to float on or in or travel on or through water or air.

Section 5
Endorsements

Endorsement 1. Subject to the terms exceptions limits and conditions of this Policy the Company will indemnify the Insured in respect of liability assumed by the Insured under any contract or agreement entered into with any Principal but only in respect of bodily injury to any person and/or loss of or damage to property happening in connection with any work or contract carried out by the Insured in connection with the Business for such Principal.

Provided: always that the Company shall have the conduct and control of all claims for which the Company may be liable by virtue of this Endorsement.

Endorsement 2. In respect of any contract for work carried out by the Insured for any person (hereinafter called 'the Principal') under which the Insured is required to effect insurance in the name of the Principal the Company will subject to the terms exceptions limits and conditions of this Policy indemnify the Principal but only so far as concerns legal liability to pay damages for bodily injury to any person and/or loss of or damage to property.

Provided that:
1. The claim is such that if made upon the Insured the Insured would be entitled to indemnity under this Policy.
2. This extension shall not apply to any greater extent than the contract entered into by the Insured necessitates.
3. The Company shall have the conduct and control of all claims for which the Principal seeks indemnity hereunder or from the Insured.

Section 6
The Schedule

The Insured:	Name Address
The Business:	
Period of Insurance:	1. from the _____ to the _____ (*both dates inclusive*) 2. Any subsequent annual period for which the Insured shall pay and the Company shall agree to accept a renewal premium.
Premiums:	£_____ First £_____ Renewal £_____ Minimum
Limit of Indemnity:	£_____
Injury:	'Bodily injury to any person' shall mean death or illness of or bodily injury to any person.
The Territorial Limits:	Great Britain the Republic of Ireland Northern Ireland the Channel Islands or the Isle of Man but in respect of business journeys and the like (excluding the supervision or execution of any work or contract) anywhere in the World.

APPENDIX 1

Signature: Signed this _____ day of _____ 19___
Checked_____

Appendix 2. Contractor's employer's liability policy

Comments on the following specimen employer's liability policy are made in Chapter 7. The numbered section headings have been added to facilitate cross-referencing to Chapter 7, and would not appear in the policy.

Section 1
Whereas the insured carrying on the business described in the Schedule and no other for the purposes of this insurance by a proposal and declaration which shall be the basis of this contract and is deemed to be incorporated herein has applied to the company (hereinafter called 'the company') for the insurance hereinafter contained and has paid or agreed to pay the Premium as consideration for such insurance.

Section 2
It is hereby agreed that if any person under a contract of service or apprenticeship with the Insured (hereinafter called 'an Employee') shall while employed in or temporarily outside Great Britain, Northern Ireland, the Isle of Man or the Channel Islands sustain bodily injury or disease caused during the Period of Insurance and arising out of and in the course of his employment by the Insured in the Business.

Section 3
The Company will subject to the terms and conditions contained herein or endorsed hereon

1. indemnify the Insured against liability at law for damages and claimant's costs and expenses in respect of such bodily injury or disease.
   Provided that
   in respect of bodily injury or disease sustained by an Employee while temporarily employed outside Great Britain, Northern Ireland, the Isle of Man or the Channel Islands the action for damages is brought against the Insured in a Court of Law in Great

APPENDIX 2

Britain, Northern Ireland, the Isle of Man or the Channel Islands.
2. where any contract or agreement entered into by the Insured with any party (hereinafter called 'the Principal') so requires:
   (a) Indemnify the Insured against liability arising in connection with and assumed by the Insured by virtue of such contract or agreement; or
   (b) indemnify the Principal in like manner to the Insured in respect of the Principal's liability arising from such contract or agreement but only so far as concerns liability as defined in this Policy to an Employee of the Insured.

   Provided that
   (c) the company shall not be liable in respect of any liability directly or indirectly caused by or contributed to by or arising from: ionising radiations or contamination by radioactivity from any nuclear fuel; or from any nuclear waste from the combustion of nuclear fuel; or the radioactive toxic explosive nuclear assembly or nuclear component thereof.
   (d) the Insured shall have arranged with the Principal for the conduct and control of all claims to be vested in the Company.
   (e) the Principal shall as though he were the Insured observe fulfil and be subject to the terms and conditions of this Policy in so far as they can apply.

Section 4
The Company will also
3. pay the Solicitor's fee incurred with its consent for representation at any Coroner's Inquest or Fatal Accident Enquiry in respect of any death.

4. pay all costs and expenses incurred with its written consent.

5. pay the Solicitor's fee incurred with its written consent for representation of the Insured at proceedings in any Court of Summary Jurisdiction arising out of any alleged breach of a statutory duty resulting in bodily injury or disease which may be the subject of indemnity under this Policy.

The indemnity provided by this Policy is deemed to be in accordance with the provisions of any law relating to compulsory insurance of liability to employees in Great Britain, Northern Ireland, the Channel Islands or the Isle of Man. The Insured shall repay to the Company all sums paid by the Company which the Company would not have been liable to pay but for the provisions of such law.

CIVIL ENGINEERING INSURANCE AND BONDING

Section 5
The Schedule

Policy No:	_____ Renewal Date _____
	Branch _____

The Insured:	Name
	Address

The Business:

Period of Insurance:	1. From the _____ to the _____
	2. Any subsequent period for which the insured shall pay and the Company shall agree to accept a renewal premium.

Premiums:	£_____ £_____ £_____
	First* Renewal* Minimum*
	* Subject to adjustment in terms of Condition 5. Estimated amount of Wages, Salaries and other Earnings.

Signature:	Signed this ____ day of _____ 19___ Checked ____

Section 6
Conditions
1. This Policy and the Schedule shall be read together as one contract and any word or expression to which a specific meaning has been attached in any part of this Policy or of the Schedule shall bear such specific meaning wherever it may appear.
2. The Insured shall give notice in writing to the Company as soon as possible after the occurrence of any accident or disease (to which this Policy relates) with full particulars thereof. Every letter claim

APPENDIX 2

writ summons and/or process shall be notified or forwarded to the Company immediately on receipt. Notice shall also be given in writing to the Company immediately the Insured shall have knowledge of any impending prosecution or inquest in connection with any accident or disease for which there may be liability under this Policy.

3. No admission offer promise payment or indemnity shall be made or given by or on behalf of the insured without the written consent of the Company which shall be entitled if it so desires to take over and conduct in the name of the Insured the defence or settlement of any claim or to prosecute in the name of the Insured for its own benefit any claim for indemnity or damages or otherwise and shall have full discretion in the conduct of any proceedings and in the settlement of any claim and the Insured shall give all such information and assistance as the Company may require.

4. The Insured shall take reasonable precautions to prevent accidents and disease.

5. The First Premium and all Renewal Premiums that may be accepted are to be regulated by the amount of Wages and Salaries and other Earnings paid to employees by the Insured during each Period of Insurance. The name of every employee and the amount of Wages Salary and other Earnings paid to him shall be duly recorded in a proper wages book. The Insured shall at all times allow the Company to inspect such books and shall supply the Company with a correct account of all such Wages Salaries and other Earnings paid during any Period of Insurance within one month from the expiry of such Period of Insurance and if the total amount so paid shall differ from the amount on which premium has been paid the difference in premium shall be met by a further proportionate payment to the Company or by a refund by the Company as the case may be.

6. The Company may cancel this Policy by sending 30 days' notice by registered letter to the Insured at his last known address and in such event the premium shall be adjusted in accordance with Condition 5.

7. The due observance and fulfilment of the terms conditions and endorsements of this Policy in so far as they relate to anything to be done or complied with by the Insured and the truth of the statements and answers in the said proposal shall be conditions precedent to any liability of the Company to make any payment under this Policy.

Appendix 3. Contract works policy

Comments on the following specimen contract works policy are made in Chapter 10. The numbered section headings have been added to facilitate cross-referencing to Chapter 10, and would not appear in the policy.

Section 1
Damage to the Works
In consideration of the Insured having paid or agreed to pay the premium the Company will indemnify the Insured in respect of loss or damage however caused to the Property Insured occurring during any period of Insurance on or adjacent to the site of any Contract within Great Britain, Ireland, Northern Ireland, the Channel Islands or the Isle of Man during any period of insurance.

Section 2
Extensions
This Policy extends to include:

1. Transit. Loss of or damage to the Property insured whilst in transit within Great Britain, Ireland, Northern Ireland, Channel Islands or the Isle of Man other than:
(a) by sea or air;
(b) any mechanically propelled vehicle under its own power;
(c) employees' tools and personal effects.

2. Principals. Any Principal as an Insured in respect of any Contract to which this Policy applies but only to the extent to which that interest is required to be insured jointly with that of the Insured by the terms of any contract entered into between the Principal and the Insured.

3. Architects' Surveyors' and Consulting Engineers' Fees. Architects' Surveyors' Consulting Engineers' and other such professional fees necessarily incurred on the reinstatement of the Property Insured consequent upon its loss or damage but not for preparing any claim.

APPENDIX 3

4. *Removal of Debris* Costs and expenses incurred by the Insured with the consent of the Company in:
(a) removing debris;
(b) dismantling and/or demolishing;
(c) shoring up or propping;
of the portion or portions of property lost or damaged which is the subject of indemnity under this Policy.

5. *Off-site Storage.* Materials and goods whilst not on the site of any contract but intended for inclusion in any Contract Works covered by this Policy where the Contractor is responsible under:
(a) clauses 14 and 30(2A) of the JCT Standard Form of Building Contract; or 54(3) of the ICE Conditions of Contract or
(b) any other standard printed contract condtions provided that the value of such materials and goods has been included in an interim certificate and the materials and goods are separately stored and identified as being designated for incorporation in a specific contract.

Section 3
Exceptions
The Company shall not be liable in respect of:

1. The amounts specified as the Insured's Retained Liability.
2. Loss of or damage to any part of the Property Insured after such property has been completed pending sale or leasing.
3. Loss of or damage to any part of the permanent works:
   (a) after such part has been completed and delivered up to the owner tenant or occupier; or
   (b) after such part has been taken into use by the owner tenant or occupier; or
   (c) for which a certificate of completion has been issued; other than where loss or damage (not otherwise excluded by this Policy) is the responsibility of the Contractor under the terms of any Maintenance Period or Defects Liability Period clause incorporated in any standard printed form of contract conditions the period of which does not exceed 12 months duration.
4. Loss of or damage to:
   (a) property forming or which has formed part of any structure prior to the commencement of the Contract or Works;
   (b) deeds bonds bills of exchange promissory notes cash bank notes cheques securities for money or stamps;
   (c) any mechanically propelled vehicle including any trailer attached

thereto licensed for road use and for which a Certificate of Motor Insurance is required other than vehicles used solely as a tool of trade on a site to which this Policy applies;
- (d) constructional plant due to its own breakdown or its own explosion but this exception shall not apply to loss of or damage to other Property Insured;
- (e) any vessel or craft made or intended to float on or in or travel on or through water or air;
- (f) property for which the Contractor is relieved of responsibility by conditions of contract.

5. The cost of:
   - (a) repairing replacing or rectifying property which is defective in material or workmanship;
   - (b) normal upkeep or normal making good.

6. Loss or damage due to:
   - (a) wear and tear rust mildew or other deterioration;
   - (b) fault defect error or omission in design plan or specification;
   - (c) confiscation nationalisation or requisition or destruction by or under the order of any government or public or local authority.

7. Loss of property by disappearance or shortage which is only revealed when an inventory is made or is not traceable to an event.

8. Penalties under contract for delay or non-completion or consequential loss or damage of any kind or description.

9. Loss or destruction of or damage to any property in Northern Ireland or loss resulting therefrom caused by or happening through or in consequence of:
   - (a) civil commotion;
   - (b) any unlawful wanton or malicious act committed maliciously by a person or persons acting on behalf of or in connection with any unlawful association.

   Note: 'Unlawful association' means any organisation which is engaged in terrorism and includes an organisation which at any relevant time is a proscribed organisation within the meaning of the Northern Ireland (Emergency Provisions) Act 1973.

   'Terrorism' means the use of violence for political ends and includes any use of violence for the purpose of putting the public or any section of the public in fear.

In any action suit or other proceedings where the Company alleges that by reason of the provision of this endorsement any loss destruction or damage is not covered by this Policy the burden of proving that such loss destruction or damage is covered shall be upon the Insured.

APPENDIX 3

This overriding exception applies to this Policy and to any extensions thereof whether such extensions be issued before or after this overriding exception except if an extension be issued hereafter which expressly cancels this overriding exception.

10. Any consequence of war invasion act of foreign enemy hostilities (whether war be declared or not) civil war rebellion revolution insurrection or military or usurped power.

11. Loss of or damage to any property whatsoever or any loss or expense whatsoever resulting or arising therefrom or any consequential loss directly caused by or contributed to by or arising from:
    (*a*) ionising radiations or contamination by radioactivity from any nuclear fuel or from any waste from the combustion of nuclear fuel;
    (*b*) the radioactive toxic explosive or other hazardous properties of any explosive nuclear assembly or nuclear component thereof.

12. Loss or damage directly occasioned by pressure waves caused by aircraft and other aerial devices travelling at sonic or supersonic speeds.

Section 4
Limits of Liability: The Liability of the Company
1. Under Item 1 and any extension hereof shall not exceed 120% of the estimated original contract price including the value of Free Materials in respect of any one Contract or Works.

2. Under Items 2, 3 and 5 shall not exceed the Sum Insured thereon.

3. Under Item 4 shall not exceed the Sum Insured thereon in respect of any one item nor 200% of that Sum Insured in respect of any one incident except as far as reinstatement thereof is made as herein stated.

In consideration of the Sum Insured not being reduced by the amount of any loss the Insured shall pay the appropriate additional premium on the amount of the loss from the date thereof to the date of the expiry of the Period of Insurance which additional premium shall be disregarded for the purpose of any adjustment of the premium under General Condition 5 of this Policy.

Definition
For the purpose of this insurance Free Materials shall be any materials supplied by or provided to the Insured for inclusion in the Contract or Works for which the Insured is responsible the value of which will not be included in the final valuation of the Works carried out or Final Contract Price and which are not otherwise excluded from this Policy.

Section 5
General conditions
1. This Policy and the Schedule shall be read together as one contract and any word or expression to which a specific meaning has been attached in any part of this Policy or of the Schedule shall bear such meaning wherever it may appear.
2. The Insured shall take all reasonable precautions to prevent loss or damage and the Company's representatives shall have access at all reasonable times to the site of any Contract or Works and the Property Insured.
3. If any change shall occur materially varying the facts existing at the commencement of the Period of Insurance or if any defects or conditions of working which render the risk more than usually hazardous are discovered the Insured shall forthwith notify the Company and in the meantime cause such additional precautions to be taken as circumstances may require.
4. The due observance and fulfilment of the terms provisions conditions and endorsements of this Policy by the Insured in so far as they relate to anything to be done or complied with by him and the truth of the statements and answers in the proposal made by the Insured (which shall be the basis of this Policy) shall be conditions precedent to any liability of the Company.
5. The premium for this Policy is provisional and has been calculated on estimates given by the Insured who shall keep accurate records containing all relevant particulars and which will be made available to the Company should they so require. The Insured shall within three months of the expiry of each Period of Insurance declare to the Company the information required and the premium for such Period of Insurance will be adjusted and a return allowed or additional premium charged as the case may be but subject to any minimum requirements.

Section 6
Claims conditions
1. (a) In the event of any occurrence which may give rise to a claim under this Policy the Insured shall as soon as possible give notice thereof to the Company in writing with full details and as far as practicable there shall not be any alteration or repair until the Company shall have had an opportunity of inspecting;
(b) In the case of theft loss or wilful damage to the Property Insured the Insured shall give notice to the police and render all

APPENDIX 3

    reasonable assistance in causing the discovery and punishment of any guilty person and in tracing and recovering such Property Insured;
    (c) In no case shall the Company be liable for any loss or damage to the Property Insured not notified to the Company within three calendar months after the event.

2. The Insured shall at the request of and at the expense of the Company do and concur in doing and permit to be done all such acts and things as may be necessary or reasonably required by the Company for the purpose of enforcing any rights and remedies or of obtaining relief or indemnity from other parties to which the Company shall be or would become entitled or subrogated upon its paying for or making good any loss or damage under this Policy whether such acts and things shall be or become necessary or required before or after their indemnification by the Company. The Insured shall not in any case be entitled to abandon any property to the Company. The Insured shall not accept any payment nor make nor accept any settlement or arrangement in respect of any loss or damage nor incur any expense in making good any loss or damage without the written consent of the Company. Any waiver of rights shall be at the expense of the Insured.

3. The Company may at its option repair reinstate or replace any property lost or damaged or pay the amount of the loss or damage in money. The Company shall not be responsible for the cost of any alterations additions improvements or overhauls carried out on the occasion of a repair.

4. If at the time of any occurrence or claim there is or but for the existence of this Policy would be any other policy of indemnity or insurance in favour of or effected by or on behalf of the Insured applicable to such occurrence or claim the Company shall not be liable under this Policy to indemnify the Insured in respect of such occurrence or claim except so far as concerns any excess beyond the amount which would be payable under such other indemnity or insurance had this Policy not been effected.

5. If any difference shall arise as to the amount to be paid under this Policy (liability being otherwise admitted) such difference shall be referred to an Arbitrator to be appointed by the parties in accordance with the Statutory provisions in that behalf for the time being in force. Where any difference is by this Condition to be referred to arbitration the making of an Award shall be a condition precedent to any right of action against the Company.

CIVIL ENGINEERING INSURANCE AND BONDING

Section 7
The Schedule

The Insured: Name
Address

Premium: £_____ Subject to General Condition 5.

The Business:

Period of Insurance:
1. From the _____ to the _____
2. Any subsequent period for which the Insured shall pay and the Company shall agree to accept a Renewal Premium.

The Contract or Works:

Any contracts or works undertaken by the Insured in the course of The Business where the estimated original contract price plus the value of Free Materials does not exceed:

£_____

The Company agrees to provide insurance in respect of the Property Insured in connection with any contract the value of which exceeds the above amount subject to the Company being given details of the contract prior to the inception of the insurance for any contract and to the acceptance by the Insured of the terms to be applied to such contracts.

The Property Insured:

*See limits of liability

Item 1
Works, temporary works and materials for use in connection therewith for which the Insured is responsible (other than property described in items 2, 3, 4 and 5).

*Sum Insured

APPENDIX 3

Item 2
Constructional plant scaffolding £
tools and equipment owned by the
Insured or for which he is responsible and not otherwise insured
excluding:
(a) any item of property which
exceeds

£_____ in value;

(b) property described in items 3, 4
and 5 below.

Item 3
Site huts and temporary buildings £
and contents thereof (other than
property described in items 2, 4 and
5) owned by the Insured or for
which he is responsible excluding
any item exceeding

£_____ in value.

Item 4
Property hired in for which the £
Insured is responsible.

Item 5
Employees' tools and personal £
effects (not being motor vehicles
gold or silver articles watches
jewellery or money) for which the
Insured is responsible and for an
amount not exceeding £150 any one
employee.

The Insured's Retained Liability:

1. In respect of items 1, 2, 3 and 4:
   (a) the first £_____ of each and every occurrence of loss or damage by storm tempest water frost subsidence landslip or collapse;
   (b) the first £_____ of each and every occurrence of loss or damage by theft or malicious persons;
   (c) the first £_____ of each and every occurrence of other loss or damage.

113

2. In respect of item 5 the first £_____ of each and every occurrence of loss or damage. In the event of any occurrence of loss or damage being subject to more than one Retained Liability the higher shall apply.

Appendix 4. Form of Bond

Comments on the following specimen bond are made in Chapter 13.

BY THIS BOND [1]We ....
of .... in the
County of .... ²We .... Limited
whose registered office is at .... in the
County of .... ³We ....
and .... carrying on business in partnership under
the name or style of ....
at .... in the
County of .... (hereinafter called 'the Contractor') [4]and ....
in the County of .... and ....
of .... in the County of ....
.... and [5]....................... Limited
whose registered office is at .... in the
County of .... (hereinafter called 'the [4]Sureties/Surety') are held

---

1 Is appropriate to an individual,
2 to a Limited Company and
3 to a Firm.
Strike out whichever two are inappropriate.
4 Is appropriate where there are two individual Sureties,
5 where the Surety is a Bank or Insurance Company.
Strike out whichever is inappropriate.

and firmly bound unto .... (hereinafter called 'the Employer') in the sum of .... pounds (£.............) for the payment of which sum the Contractor and the ⁴Sureties/Surety bind themselves their successors and assigns jointly and severally by these presents.

Sealed with our respective seals and dated this ...... day of ...... 19......

WHEREAS the Contractor by an Agreement made between the Employer of the one part and the Contractor of the other part has entered into a Contract (hereinafter called 'the said Contract') for the construction and completion of the Works and maintenance of the Permanent Works as therein mentioned in conformity with the provisions of the said Contract.

NOW THE CONDITIONS of the above-written Bond are such that if:
(a) The Contractor shall subject to Condition (c) hereof duly perform and observe all the terms provisions conditions and stipulations of the said Contract on the Contractor's part to be performed and observed according to the true purport intent and meaning thereof or if
(b) on default by the Contractor the Sureties/Surety shall satisfy and discharge the damages sustained by the Employer thereby up to the amount of the above-written Bond or if
(c) the Engineer named in Clause 1 of the said Contract shall pursuant to the provisions of Clause 61 thereof issue a Maintenance Certificate then upon the date stated therein (hereinafter called 'the Relevant Date')

this obligation shall be null and void but otherwise shall remain in full force and effect but no alteration in the terms of the said Contract made by agreement between the Employer and the Contractor or in the extent or nature of the Works to be constructed completed and maintained thereunder and no allowance of time by the Employer or the Engineer under the said Contract nor any forbearance or forgiveness in or in respect of any matter or thing concerning the said Contract on the part of the Employer or the said Engineer shall in any way release the Sureties/Surety from any liability under the above-written Bond.

PROVIDED ALWAYS that if any dispute or difference shall arise between the Employer and the Contractor concerning the Relevant Date or otherwise as to the withholding of the Maintenance Certificate then

APPENDIX 4

for the purposes of this Bond only and without prejudice to the resolution or determination pursuant to the provisions of the said Contract of any dispute or difference whatsoever between the Employer and Contractor the Relevant Date shall be such as may be:

(*a*) agreed in writing between the Employer and the Contractor or

(*b*) if either the Employer or the Contractor shall be aggrieved at the date stated in the said Maintenance Certificate or otherwise as to the issue or withholding of the said Maintenance Certificate the party so aggrieved shall forthwith by notice in writing to the other refer any such dispute or difference to the arbitration of a person to be agreed upon between the parties or (if the parties fail to appoint an arbitrator within one calendar month of the service of the notice as aforesaid) a person to be appointed on the application of either party by the President for the time being of the Institution of Civil Engineers and such arbitrator shall forthwith and with all due expedition enter upon the reference and make an award thereon which award shall be final and conclusive to determine the Relevant Date for the purposes of this Bond. If the arbitrator declines the appointment or after appointment is removed by order of a competent court or is incapable of acting or dies and the parties do not within one calendar month of the vacancy arising fill the vacancy then the President for the time being of the Institution of Civil Engineers may on the application of either party appoint an arbitrator to fill the vacancy. in any case where the President for the time being of the Institution of Civil Engineers is not able to exercise the aforesaid functions conferred upon him the said functions may be exercised on his behalf by a Vice-President for the time being of the said Institution.

Signed Sealed and Delivered by the said
  in the presence of:

The common Seal of
                        Limited
was hereunto affixed in the presence of:

(*Similar forms of Attestation Clause for the Sureties or Surety*)

Appendix 5. Professional indemnity insurance for consulting engineers

Comments on the following specimen professional indemnity insurance policy are made in Chapter 15.

WHEREAS the Insured named in the Schedule have made a written proposal, bearing the date stated in the Schedule containing particulars and statements which it is hereby agreed are the basis of this Policy and are to be considered as incorporated herein, and have paid the premium stated in the Schedule.

NOW We, the Insurers, to the extent and in the manner hereinafter provided, hereby agree to indemnify the Insured against claims made against the Insured during the period of insurance specified in the Schedule in respect of any legal liability or alleged legal liability arising out of the Insured's activities in the conduct of the practice(s) as Consulting Engineers due to any negligent act error or omission by the insured or his employees.

THE LIABILITY of the Insurers shall not exceed the Limits of Liability stated in the Schedule in respect of each claim, or series of claims, arising out of one source or cause, under this insurance except that the Insurers agree to pay, in addition to the limits of liability, all costs and expenses incurred with their written consent in the investigation, defence or settlement of any claim under this insurance.

IN THE EVENT that any claim exceeds the indemnity available under this Policy the Insurers hereon shall not be liable for the total costs and expenses in respect of that claim but shall contribute to the total costs and expense in the proportion that the indemnity paid under this Policy in respect of that claim bears to the total indemnity paid by this policy and any Policies effected in excess hereof.

HOWEVER, the Insurers shall only be liable in respect of any claim under this insurance for that part of the claim, or that part of the aggregate of a series of claims arising out of one source or cause, which exceeds the amount stated in the Schedule as 'the Excess'.

Exceptions

The Insurers shall not be liable in respect of:

1. any loss brought about or contributed to by the dishonesty of any of the Insured's partners or directors.
2. any claim arising from the loss of money or securities by theft, embezzlement or misappropriation by any of the Insured's employees.
3. any legal liability of whatsoever nature directly or indirectly caused by or contributed to by or arising from:
   (a) ionising radiations or contamination by radioactivity from any nuclear fuel or from any nuclear waste from the combustion of nuclear fuel;
   (b) the radioactive toxic explosive or other hazardous properties of any explosive nuclear assembly or nuclear component thereof.

Memoranda

Memo 1. Project managers' extension

This Insurance is extended to include the Insured's activities as Project Managers.

Provided always that this extension shall only cover the Insured for their legal liability for negligent acts, errors or omissions as managers of a project where they are remunerated for a fee for their services and shall be limited to the overall control and general supervision of the contract. This extension shall not however cover claims made against the Insured for

(a) any claim resulting from failure to procure or maintain any financing for the payment of contract work or services in connection therewith from any cause whatsoever
(b) any claim which would normally be the responsibility of the building contractor if a separate Project Manager were not appointed
(c) any claim as a result of failure to effect and/or maintain insurance
(d) insolvency of any of the parties involved in the project
(e) any liability assumed by the Insured under contract which would not have attached to them but for the existence of the contract
(f) error or omission by the Insured in estimates of probable construction cost or cost estimates being exceeded.

All other terms and conditions remain unaltered.

Memo 2. Libel and slander extension
Subject otherwise to the terms, clauses and conditions of this Policy, the Insurers will indemnify the Insured for all sums which the Insured may become legally liable to pay in respect of claims made upon them in direct consequence of any libel or slander uttered by the Insured in their professional capacity, provided that this insurance shall not extend to any matter contained in a journal or publication or in any communication or contribution to the Press, Radio or Television.

Memo 3. Dishonesty of employees extension
Subject otherwise to the terms, clauses and conditions of this Policy, the Insurers will indemnify the Insured for any claim brought about, or contributed to, by the dishonest, fraudulent, criminal or malicious act or omission of any person at any time employed by the Insured.

Memo 4. Loss of documents extension
Notwithstanding anything contained herein to the contrary, it is understood and agreed that if during the currency of this Policy the Insured shall discover, and shall within seven days of the date of discovery give written notice thereof to the Insurers, that any Documents (as hereinafter defined) the property of, or entrusted to, the Insured which now or hereafter are or are by them supposed or believed to be in their hands or in the hands of any other Party or Parties to or with whom such Documents have been entrusted, lodged or deposited by the Assured in the ordinary course of business have been destroyed or damaged or lost or mislaid and after diligent search cannot be found this Policy shall indemnify the Insured for

(a) any liabilities of whatsoever nature which they may incur to Third Parties in consequence of such Documents having been so destroyed, damaged, lost or mislaid.
(b) all costs, charges and expenses of whatsoever nature incurred by the insured in replacing and/or restoring such Documents.

Provided always that
1. the amount of any claim for costs and expenses as above shall be supported by Bills, and/or Accounts which shall be subject to approval by some competent person to be nominated by the Insurers with the approval of the Insured.
2. no liability shall attach hereto for any loss brought about or contributed to by the dishonesty of any of the Insured's partners.
3. on payment of any loss hereunder, the Insured shall subrogate to insurers their right of procedure against any other person or persons for the recovery thereof.

APPENDIX 5

4. Insurers' liability in respect of any and all losses hereunder shall be limited to the sum of £10,000.
5. the Excess as stated in the Schedule of the Policy wording shall not apply to this Loss of Documents Extension.
6. 'Documents' shall mean Deeds, Wills, Agreements, Maps, Plans, Records, written or printed Books, Letters, Certificates or written or printed Documents and/or Forms of any nature whatsoever (excluding, however, any Bearer Bonds or Coupons, Bank or Currency Notes or other negotiable paper) used in connection with the Insured's business.
7. It is understood and agreed that the limit of liability under this Loss of Documents Extension is in addition to the limit of liability as stated in the Schedule.
8. For the purposes of this extension and no other the Excess is £50 for each and every claim.

Memo 5
Subject otherwise to the terms, clauses and conditions of this Policy the Insurers will indemnify the past, present or future partners and employees of the Assured as co-assured, provided that such partners and employees shall comply with the terms and conditions of this Policy in so far as they can apply.

Memo 6
The Insurers hereby agree to pay all costs incurred by the Insured in connection with legal proceedings taken by the Insured for the recovery of professional fees due in accordance with the Scale of Professional Charges sanctioned by the Association of Consulting Engineers provided that:

(a) no claim under this Clause shall attach unless and until
  (i) the Insured has instituted proceedings for recovery and
  (ii) the party sued has intimated his intention to raise by way of answer a counter-claim such as would be covered under this insurance.
(b) the Insurers shall be advised of any intended claim immediately the grounds of the defence have been disclosed to the Insured.

Memo 7
It is noted and agreed that the term 'employee' as used in the Policy shall be deemed to include 'freelance employees' engaged by the Insured.

Memo 8
It is noted and agreed that this insurance shall indemnify the Insured

against liability arising out of work performed by sub-contractors on the Insured's behalf. However, the insurers shall retain their full rights of subrogation in respect thereof.

Memo 9
It is noted and agreed that subject otherwise to the terms, exclusions, limitations and conditions of the Policy it is hereby understood and agreed that the practice(s) as Consulting Engineers referred to in the Insuring Clause shall be deemed to include the commissioning of installations, including physical work done by the Insured in connection therewith, and the preparation of planned maintenance documentation, manuals and schedules.

Conditions
1. Immediate notice shall be given by the Insured in writing to the Insurers of

    (i) any claim or intimation to the Insured of possible claim made against the Insured which may give rise to a claim under this Policy.
    (ii) any occurrence or circumstance, of which the Insured shall become aware during the subsistence of this Policy, which may subsequently give rise to a claim against the Insured. Such notice having been given, any claim, to which that occurrence or circumstance has given rise, which may be made after the expiration of the period specified in the Schedule shall be deemed to be made during the subsistence of this Policy for the purposes of this insurance.

   Furthermore the Insured shall upon request give to the Insurers all such information and assistance as the Insurers may reasonably require and as may be in the Insured's power and will in all such matters do and concur in doing all such things as the Insurers may require.

2. The Insured shall not be required to contest any legal proceedings unless a Queen's Counsel (to be mutually agreed upon) shall advise that such claim could be contested with a reasonable prospect of success and the Insured consents thereto such consent not to be unreasonably withheld. In the event of a difference of opinion between the Insured and the Insurers as to what constitutes an unreasonable refusal the Chairman for the time being of the Association of Consulting Engineers shall nominate an independent referee to decide this point only and the decision of such referee shall be binding on the Insured and the Insurers.

APPENDIX 5

3. All differences arising out of this Policy shall be referred to the arbitration of some person to be appointed by both parties or if they cannot agree upon a single arbitrator to the decision of two arbitrators one to be appointed in writing by each party and in case of disagreement between the arbitrators to the decision of an umpire who shall have been appointed in writing by the arbitrators before entering on the reference and an award shall be a condition precedent to any liability of the Insurers or any right of action against the Insurers. The provisions of the Arbitration Acts shall apply to any such arbitration and the place of arbitration shall be London England.

4. If the Insurers shall disclaim liability to the Insured in respect of any claim made upon the Insured the Insured shall be at liberty without prejudice to his claim for indemnity hereunder to settle or compromise such claims or submit to any judgement or arbitration award in respect thereof and the sum payable by the Insured in respect of any such settlement compromise judgement or arbitration award shall be accepted by the Insurers as the amount payable by the Insurers to the Insured subject to the terms of this Policy and subject to the liability of the Insurers being established by arbitration as herein provided.

5. The due observance and fulfilment of the terms of this policy so far as they relate to anything to be done or complied with by the Insured and the truth of any proposal shall be conditions precedent to any liability of the Insurers to make any payment under this Policy.

6. If any payment is made under this insurance in respect of a claim the Insurers are thereupon subrogated to all the Insured's rights of recovery in relation thereto. However, the Insurers shall not exercise any such rights against any employee of the Insured unless the claim has been brought about or contributed to by the dishonest, fraudulent, criminal or malicious act or omission of the employee.

7. If at the time any claim arises under this Insurance the Insured is or would but for the existence of this Insurance be entitled to indemnity under any other policy or policies by the Insurers shall not be liable except in respect of any excess beyond the amount which would have been payable under such other policy or policies had this Insurance not been effected.

8. If the Insured shall make any claim knowing the same to be false or fraudulent as regards amount or otherwise this Insurance shall become void and all claim hereunder shall be forfeited.

1. THE ASSURED:

2. ADDRESS OF THE ASSURED:

3. THE PERIOD OF INSURANCE:

4. THE INDEMNITY:

5. THE EXCESS:

6. THE PREMIUM:

7. DATE OF PROPOSAL FORM:

8. TERRITORIAL LIMITS:

9. PROFESSIONAL ACTIVITIES AND DUTIES:

Appendix 6. Professional indemnity insurance for engineering contractors' design activities

Comments on the following specimen professional indemnity insurance policy are made in Chapter 15.

WHEREAS the Assured has made a written proposal to Us, the Underwriters, which proposal together with other information supplied by the Assured, shall be deemed to be incorporated herein and shall be the basis of this contract and has paid to us the premium stated in the Schedule.

NOW THEREFORE, We, the Underwriters jointly agree to indemnify the Assured up to but not exceeding the indemnity stated in the Schedule for any sum or sums which the Assured may become legally liable to pay arising from any claim or claims made against them during the period stated in the Schedule as a result of any negligent act, error and/or omission in the conduct and execution of all their professional activities and duties.

Provided always that the Underwriters' liability in respect of all claims shall not exceed the indemnity stated in the Schedule.

FURTHER, it is understood and agreed that the Underwriters will pay in addition to the indemnity stated in the Schedule costs and expenses incurred with the Underwriters' written consent in the defence and/or settlement of any claim. However, if a payment in excess of the amount of indemnity available under this insurance has to be made to dispose of a claim, the Underwriters' liability in respect of such costs and expenses incurred shall be such as the amount of the indemnity available under this insurance bears to the total amount paid to dispose of the claim.

The excess
PROVIDED ALWAYS THAT the Underwriters shall be liable only, in respect of any one claim, for that part of the claim which exceeds the relevant excess stated in the Schedule. It being understood and agreed that if any expenditure is incurred by the Underwriters which, by virtue

of this clause, is the responsibility of the Assured then such amount shall be reimbursed to the Underwriters by the Assured forthwith on demand.

Exclusions
1. This Policy does not cover any claim or claims arising out of:
    (i) Bodily injury sustained by any person arising out of and in the course of his employment by the Assured under a contract of service or apprenticeship with the Assured.
    (ii) Any neglect, error and/or omission to effect or maintain insurance and/or to provide finance or advice on financial matters.
    (iii) Any claim against the Assured for libel and/or slander and/or as a result of any dishonest, malicious, criminal or illegal acts of the Assured.
    (iv) The insolvency of the Assured.
    (v) The ownership, use and/or occupation of property mobile and/or immobile including waterborne vessel or craft or aircraft or motor vehicles by, to or on behalf of the Assured.
    (vi) The cost of replacing documents which have been lost, mislaid or destroyed.
    (vii) Any legal liability of whatsoever nature directly or indirectly caused by or contributed to by or arising from

      (a) ionising radiations or contamination by radioactivity from any nuclear fuel or from any waste from the combustion of nuclear fuel.
      (b) the radioactive, toxic, explosive or other hazardous properties of any explosive nuclear assembly or nuclear component thereof.
2. This Policy shall not indemnify the Assured against liability for any fines, penalties, punitive or exemplary damages and/or liquidated damages.

Special provisions
1. Where the Assured incur loss or expenses in remedying defects in any contract works and arising from any negligent act, error and/or omission in connection with their professional activities and duties, the Underwriters agree to consider such loss or expense as a claim against the Assured.
Provided always that

    (a) such defect is discovered prior to handover of such contract works to the Assured's employer,

APPENDIX 6

(b) the burden of proving any claim under this Special Provision shall be upon the Assured,

(c) there shall be excluded from any indemnity under this Special Provision any amount recoverable or which would be recoverable but for this exclusion under any other insurance policy.

Conditions
1. *Assured's duties in the event of claim*
   (i) The Assured shall as a condition precedent to their right to be indemnified under this insurance give immediate written notice to the person(s) named for that purpose in the Schedule for transmission to the Underwriters

   (a) of any claim made against them,
   (b) of the receipt of notice from any person of an intention to make a claim against them,
   (c) of any defect in any contract works following which loss or expense may be incurred by the Assured in accordance with Special Provision 1 (One),
   (d) of any circumstance of which they shall become aware during the period stated in the Schedule, which is likely to give rise to a claim against them. Such notice having been given, any claim, to which that circumstance has given rise, which may be made after the expiration of the period stated in the Schedule shall be deemed for the purpose of this Policy to have been made during the period stated in the Schedule.

   (ii) The Assured shall not admit liability or settle or make or promise any payment in respect of any claim which may be the subject of indemnity hereunder or incur any costs or expenses in connection therewith without the written consent of the Underwriters who if they so wish shall be entitled to take over and conduct in the name of the Assured the defence and/or settlement of any such claim for which purpose the Assured shall give all such information and assistance as the Underwriters may reasonably require.

2. *Litigation*
   The Underwriters will not require the Assured to dispute any claim unless a Queen's Counsel or lawyer of comparable standing in the territory concerned (to be mutually agreed upon by the Underwriters and the Assured) advise that the same could be contested

with a reasonable prospect of success by the Assured and the Assured consents to such claim being contested, such consent not to be unreasonably withheld. In the event of any dispute arising between the Assured and the Underwriters as to what constitutes an unreasonable refusal to contest a claim at Law, the President for the time being of the Chartered Institute of Arbitrators shall nominate a Referee to decide this point (only) and the decision of such Referee shall be binding on both parties.

3. *Subrogation*
It is hereby agreed that if any payment is made under this insurance in respect of a claim the Underwriters are thereupon subrogated to all the Assured's rights of recovery in relation thereto. However, the Underwriters shall not exercise any such rights against any employee of the Assured unless the claim has been brought about or contributed to by the dishonest, fraudulent, criminal or malicious act or omission of the employee.

4. *Fraudulent claims*
If an Assured shall make any claim knowing the same to be fraudulent or false, as regards amount or otherwise, this insurance shall become void and all claims hereunder shall be forfeited in respect of that particular Assured.

5. *Other insurance*
Insofar as it is within their power the Assured will produce information regarding any other insurance of the Assured's professional activities and duties on receipt of the Underwriters' request.

6. *Non-contribution clause*
If the liability which is the subject of a claim under this Policy is covered by any other insurance and that insurance has a Non-Contribution Clause, the Underwriters under this Policy will not pay more than their rateable proportion.

It is understood and agreed that this Clause does not apply to claims the subject matter of Memo 3 or Special Provision 1 (One).

Dishonesty of employees
It is understood and agreed, subject otherwise to the terms, exclusions, limitations and conditions contained in this Policy, that this Policy is extended to indemnify the Assured for any claim brought about, or contributed to, by the dishonest, fraudulent, criminal or malicious act or omission of any person at any time employed by the Assured.

APPENDIX 6

Infringement of copyright
It is understood and agreed, subject otherwise to the terms, exclusions, limitations and conditions contained in this Policy, that this Policy is extended to indemnify the Assured for any claim or claims made against the Assured arising out of any infringement of copyright or breach of patent.

---

1. THE ASSURED:

2. ADDRESS OF THE ASSURED:

3. THE PERIOD OF INSURANCE:

4. THE INDEMNITY:

5. THE EXCESS:

6. THE PREMIUM:

7. DATE OF PROPOSAL FORM:

8. TERRITORIAL LIMITS: Worldwide

9. PROFESSIONAL ACTIVITIES AND DUTIES
   1. Work undertaken by the Design Department
   2. Feasibility Studies

Appendix 7. Further reading

Madge Peter. *Indemnity and insurance aspects of building contracts.* Royal Institute of British Architects Publications Ltd, London, 1985.

Eaglestone F.N. and Smyth C. *Insurance under the ICE Contract.* George Godwin, London, 1985.